Fort Stanton
An Illustrated History

Legacy of Honor, Tradition of Healing...

by Lynda A. Sánchez

Crossed Saber Cap Insignia, Sam Townley Collection

Photo by David L. Tremblay

Fort Stanton
An Illustrated History

by *Lynda A. Sánchez*
with Contributions by:
Frederick W. Nolan, Jay Smith, Robert M. Utley
Photography by David L. Tremblay

Edited by Ann Buffington

Creative Design, Layout and Photo Editing by Laura Reynolds

Published by
Write Designs, LTD
Ruidoso, NM

5 4 3 2 1

Manufactured in the USA
ISBN:978-0-9772614-9-9
Library of Congress Control Number: 2010926930

Second, Revised Edition - 2010

This book is non-fiction. Reference to historical events, individuals, or actions are compiled from data drafted by the Author.

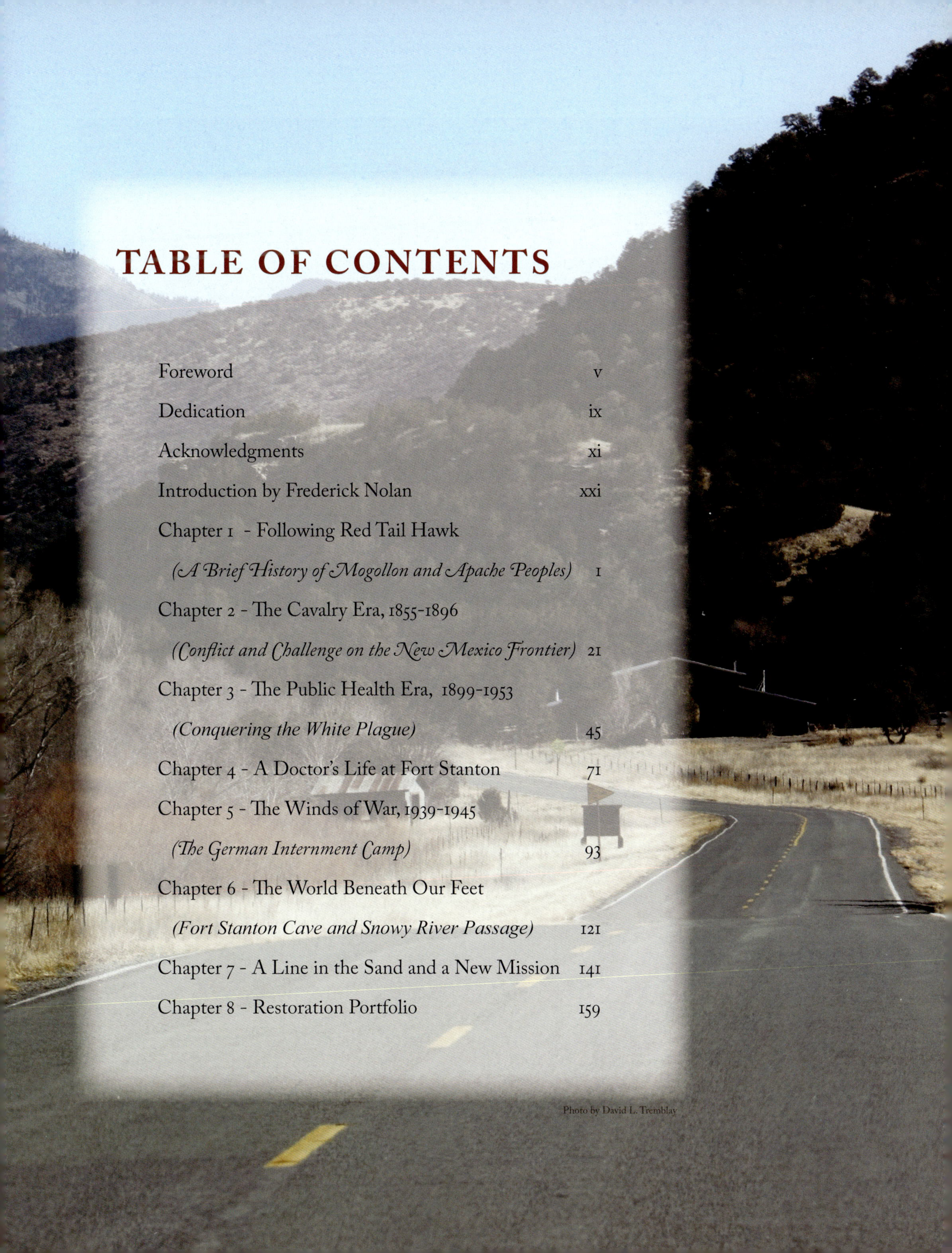

TABLE OF CONTENTS

Photo by David L. Tremblay

Photo by David L. Tremblay

Photo by David L. Tremblay

FOREWORD:

A unique partnership that has led to this publication about Fort Stanton was forged because of one specific photo depicting a red tailed hawk. I saw it in *The Ruidoso News*, called the photographer, whom I did not know, and asked for permission to use the shot at a later time. Little did I realize that the photographer was David Tremblay, a generous and talented artist in his own right. His photographs for all seasons and settings with the many creatures we have here in Lincoln County are becoming familiar to many area residents. Mostly a wild life/nature photographer, David now claims Fort Stanton as part of his repertoire.

As noted in the beginning of the acknowledgments section, this book would not have been possible without David and his many talents. He is unbelievably patient and has allowed us to expand beyond the normal illustrated histories I have seen. We include art, poetry, historic photographs and of course his incredible contemporary shots. I hope that the readers will see these attributes throughout the book and next time they see a red tailed hawk floating on the breeze or streaking across the sky, say "thanks".

Lynda A. Sánchez
Lincoln, New Mexico

The reason I became involved with the Fort is simple in hindsight. I remember the day Deedee and I delivered the hawk picture to Lynda for 'a book she might write one day'. We were to meet at the little Museum on the Fort grounds. She invited us to become part of her tour group with the Hubbard Museum staff. As we learned, Lynda has given innumerable (about 70 plus) tours to help educate the public about this wonderful place. We came away with a deep appreciation for the history of the fort, but what impressed the both of us most was her heart felt dedication to the preservation and restoration of the old place. SHE IS CONTAGIOUS! All the bits of information she passed along to us have made us feel as if we were a part of her passion.

And then one morning *(only God knows why)* I got this bright idea to help in some way with photography and boy have we ever gotten a priceless education along the way. The rest will be history…as in this book you are holding in your hands.

David L. Tremblay
Alto, New Mexico

(Opposite Page) Catholic Chapel on the Grounds of Fort Stanton

Photo by David L. Tremblay

Sieera Blanca, The Sacred Mountain
Photo by David L. Trembley

Dedication:

From my heart, (*de mi corazón*) I wish to dedicate this book to the many peoples (Anglo, Apache and Hispanic) who came before any of us ever arrived in this magnificent land dominated by the Sacred Mountain, Sierra Blanca. Their intrepid spirit, toughness and dreams of a new beginning, a family and a way of life that allowed for growth, freedom and protection of our precious natural and cultural resources is ever present as we narrate the fascinating Fort Stanton saga. May we be deserving of their trust. And to all the individuals and groups who love Fort Stanton and have come together in this special restoration project, I can only say *mil gracias* a thousand times.

THE UNION
1859

Sierra Blanca
Photo by David L. Tremblay

ACKNOWLEDGMENTS

Individuals whose years of experience and knowledge have encouraged me to tell the Fort Stanton story also have commentary with regard to Fort Stanton's future. Their contributions are included throughout this book. These special friends are historians, and business people, or directors of their respective institutions and include:

Frederick W. Nolan, Historian and author/musician extraordinaire-mi buen amigo; **Ernesto Ortega**, former National Park Service and preservation expert who now serves as Director of New Mexico State Monuments; **Jay Smith**, Director of the Hubbard Museum of the American West; **Marc Simmons**, author extraordinaire and keeper of New Mexico's history; **Robert (Bob) Utley**, former National Park Service Historian and author of numerous books about the West.

A very special and heartfelt thank you to **David Tremblay** who has assisted with creative photography techniques, scanning, design, photographs of the Fort, and general encouragement. *This book is possible because of you and the partnership we have forged to help tell the Fort Stanton story!*

I would also like to thank **my husband, James**, and **my daughter, Katherine**, for their support, encouragement and understanding of the journey I have undertaken to get to this point in time. Secondly, the people who stood by me with patience, a willingness to work, to push the

(Opposite Page) James D. Roehrig playing Taps on a rainy Memorial Day 2009 at Fort Stanton.

politicos and yes, even some of those *politicos* who really did care about the region and its history. They helped support the preservation and protection of Fort Stanton and are listed below.

While there are many who may be "interested" in the fort, the following individuals or groups specifically assisted me in my research and education about Fort Stanton history as it really was:

My Editor, **Ms. Ann Buffington**, author/historian in her own right, always had good suggestions and knowledge of the King's English.

Ms. Laura Reynolds, "go to gal" publisher and designer of the book, as well as major Fort Stanton advocate.

The "Kids" of Fort Stanton who grew up at the Fort from the 1930's through the end of the Public Health Service years. I especially want to thank **Claire Boyd Fluharty, Danna Kusianovich Henderson**, and **Tom (Skipper) Gould** for their desire to preserve that segment of the Fort's "tradition of healing" era. Many interviews and photos later I am much more aware of the treasured history and nostalgia of that epoch.

Mr. Mike Bilbo, Bureau of Land Management, cave specialist and recreation planner as well as re-enactor and historian.

Mr. Edwin Roberson, Bureau of Land Management and former Roswell Field Office Director; **Mr. Doug Burger**, current Roswell Field Office Director and his staff.

Mr. John Corcoran and the Fort Stanton Cave Study Group for their 40 years of dedicated service in mapping and exploring the world beneath our feet, and for teaching me about Fort Stanton Cave and the famed Snowy River passage.

The dozens of folks who wrote letters, helped get petitions signed, who faxed, called and sent e-mails during our "sub-division" battle.

Mr. Bill Hume, with the Governor's office, without whose help the project would have never come to fruition. Governor Richardson told him to "fix the Fort Stanton problem" and patiently working with me and others, Mr. Hume, whom I found out also loved our region's history and appreciated that heritage, was instrumental along with **Stuart Ashman**, Cabinet Secretary for Cultural Affairs and **Mr. John Garcia**, Cabinet Secretary for Veterans' Services in the designation of the site in 2007 as a State Monument.

Our area Legislators, especially **Representative Dub Williams**, Senator **Rod Adair** and **Senator Tim Jennings** and others who recognized the value of the site in the total scheme of things.

A very specific thank you to **Nick Skogland** and **Sam Townley** who have worked with chewing gum, duct tape and baling wire to hold the Fort together all these many years. Their good cheer, concern, and assistance I shall never forget.

Members of our Archaeology survey crew and most especially **Archaeologist Charles Haecker**, Project photographer, **Joe Arcure, Chris Adams, Jeannie Adams, Louise Ritch, Bob Davis, Jane Jackson, Richard Lucero, Dorothy Foyt, Trink Edmonds, Cody Browning, Holly Houghten**, and **Jim McBride** (who also recorded the story of the German Internment camp and the Public Health Service in his three books.)

Joe and Carolyn Arcure, Richard T. Bryant, David and Deedee Tremblay, (creative photographers and/or supporters all);

Ms Diane Stallings and *The Ruidoso News*; and **Jay Miller** of *"Inside the Capitol"* for all their great coverage.

There are literally dozens of docents and locals who have given of their time and expertise over those years. Unfortunately I do not know them all but of those I have had the privilege to work with and who were supporters of the Fort's preservation I would acknowledge the following who were especially helpful to me:

Ms. Charlotte Rowe, (PIC), **Mr. Clinton Smith, Mr. Tad Vanderhye** (all from the Fort Stanton Foundation Board of Directors). I would also be remiss if I did not mention in particular **Ms. Sally Burkstaller** and her struggle during the '90's, after the Fort was closed, to shake up

(Left to Right)
Trink Edmonds, Louise Ritch, Jeannie Adams.
Archaeology Team Crew who remind us of the efforts required to bring history to light.

Photo by Joseph Arcure

Santa Fe and the locals. She helped them to understand what a resource we had in Lincoln County and she provided the ground work for what was to come. Others would include **Bev and Jack Carpenter, Tony and Nisha Hoffman, Jan Sawyer, Dick and Mary Samuels, Earl Pittman, Gary Cozzens, Paul Happel, Bob and Peggy Whittemore, Bill Murry, Van Shamblin, Jerrold Flores**, along with **Ms. Barbara Heiser** and the Outryders, **Ms. Jean Fulton** and **Mr. Pat Taylor** of Cornerstones Community Partnerships; **Ms. Betty Lou Joiner** Fort Stanton's long-time Postmaster; **Mr. Martin Hibbs, Ted Allen** and **Steve Carter, James Sánchez, Kevin and Susan Hanratty**, all for their work toward the Restoration and stabilization of the Historic Stables, and **Ms. Trink Edmonds**, docent, educator and all around "she can do it" kind of person.

Special recognition must go to *True West Magazine* and their crusading and talented President and CEO, **Bob Boze Bell**, and their Managing Editor, **Meghan Saar**. They have followed the battles, and supported the preservation efforts at the fort and have now created their own Preservation section of the Magazine. They understand that if we don't help preserve the West now, what does it matter if we write about places that no longer exist, other then with regret and remorse.

Mr. Larry Holt, for his dedication to the Merchant Marine Military Cemetery and the Memorial Day Services held every year saluting our fallen Veterans from all branches of service; **Mr. Sid Alford**, former foreman of the Coe Ranch and veteran advocate;

Mr. Bob McCubbin and **Michael Vinson** for the wonderful 1896 painting of Fort Stanton displayed herein.

The staff of the Hubbard Museum including **Director Jay Smith, Patsy Jackson**, Curator of Education and Design expert **Mr. David Mandel**; Re-enactors **Ken Dusenberry, Carl Reed, Larry Pope** and **Mike Bilbo** among others for their support and their dedication to the history of our military past;

Ms. Deann Kessler, Monument Manger, Lincoln State Monument who bridged the gap as Interim manager at Fort Stanton and her dedicated staff.

Members of the Lincoln County Historical Society Board including **Ray Dean, Herb Marsh, Karen Mills, Sadie Mann,** and **Jay Smith**; **Mr. Tom Stewart**, (Colonel, U.S. Army retired) our intrepid Lincoln County Manger whose strong support and many letters were greatly appreciated. Specific members of the Lincoln County Commission also need to be included here for both their support for the Fort and against the subdivision threat including: Former Chairman, **Rick Simpson, Earl Hobbs**, and **Maury St. John**; present Chairman, **Mr. Tom Battin**, and members **Mr. Don Williams, Mr. Dave Parks, Ms. Eileen Lovelace Sedillo, Ms. Jackie Powell** and **Alan Morel**, County attorney;

Ms. Jennifer Riggs, formerly of Civigenics, and her staff for allowing me access to many of the structures and the quadrangle during the time of their program being on site;

Otero County Electric CEO, **Mr. Carroll Waggoner,** and their Board of Directors and especially **Preston Stone**, **Fred Hansen** and **Bill Bird**; for their financial support of the Archaeology survey and stables' restoration; **Ms. Mary Serna** and **Tommy Spottedbird** of the St. Joseph's Apache Mission and Veterans' Memorial Restoration Project;

Archaeologists **Brian Kenny** (Arizona) and **Karl Laumbach** (New Mexico) for commentary and that presentation at the "Pecos Conference", 2008;

Ms. Charmay Allred and author **Michael McGarrity**, Santa Fe, for their steadfast guidance and encouragement;

Mike Stevenson of both the Historical Society of New Mexico and the Museum of New Mexico Board of Regents for his support and encouragement;

Barbara and Jack Pruitt; **Tony Davis** and **Joyce Westerbur** for their time and that trip to Albuquerque! And **Pam MacArthur**, mi amiga especial.

The Lincoln County Garden Club;

Ms. Sandra Smith for her many talents and high tech skills;

I SALUTE YOU ALL!! (If I have left a name out that should be here, someone please let me know and it will be corrected in the next publication cycle of the book. Errors are mine alone.)

The opportunity is at hand and we have a true gem, an authentic Fort that has been and will continue to be a guiding light in terms of preserving our heritage and the custom and culture of the people who came before us.

I am truly blessed and thankful for everyone's guidance and participation. As we move forward toward a new dawn, Fort Stanton's "Legacy of Honor and Tradition of Healing" will emerge in a much more detailed light, yet, the legend, the conflict, the story of gutsy men and women, and the bloody, ruthless villains who even had their "moments", all converge here to engage you, the reader, in a special way. I hope that when you complete the book, and review the many images herein, that you, too, will be inspired to visit the site, appreciate the resource and in some way begin to help in the preservation of a unique and memorable example of Western Americana.

Lynda A. Sánchez
Lincoln, New Mexico
2009

Marine Hospital truck aka "Hunting Party" c.1920's Courtesy Fort Stanton "Kids'" archives

Courtesy Danna Henderson

Danna Kusianovich and Barney Aldaz enjoying a day at the Lincoln "Billy the Kid" Pageant. c 1940.

Photo by Joseph Arcure

Sam, J Joe, and Nick.

Photo by David L. Tremblay

Photo by David L. Tremblay

My Fort Stanton

by "That Englishman"
Frederick Nolan

I can still recall the disappointment I felt when, after two decades of researching the history of the West, I first crossed the wide Missouri–this was in 1971—and discovered that at nearly every site of historical significance I visited, there was hardly any trace of what had made the place significant. The house on Lafayette Street where Jesse James was assassinated in 1882 was no longer there: it had been moved to a new location to make it easier for the tourists to find, rendering it as historically valuable as a polystyrene cup. At Dodge City they had razed the remnants of the old cow town, replaced it with a parking lot, and erected a "replica" half a mile away where they sold bricks from the original buildings for a few dollars a shot. There was another replica in Abilene; in Ellsworth there was nothing much at all. There was only

(Above) Laundry Row
(Opposite) Historic Fort Stanton stables with 1930's silos in background.

a chunk of Fort Larned (with souvenir shop, of course) and at the site of Bent's Fort there was nothing except a concrete template to show where the Fort had been; that, too, now boasts a replica.

The above is so important for people to realize. We have the real deal here in Lincoln County, New Mexico!!!

In spite of all this, I thought, I hoped, things might get better when we got into New Mexico, but Santa Fe was a tourist trap and Las Vegas, with its famous Palace Hotel boarded up, was drab. At Fort Sumner, site of the world's first concentration camp, there was not so much as a notice board mentioning imprisoned Navajos and Apache, not even a flagpole marking the old parade ground, nothing but a ramshackle "museum" and a sadly neglected cemetery with Billy the Kid's grave "protected" by a chicken-wire fence. Of course, there is a fine monument there now and even some walls and blocks to indicate where the old Fort was, but fine as they are, they are still not the real thing.

So you can imagine my relief when we rolled into Lincoln. *This* was what I had traveled all those thousands of miles for. Here, the book I had written came to life: I could the ***see*** old Murphy-Dolan store, standing where it had always been on the largely unchanged street that John Henry Tunstall had once walked down. I could walk into the store he had built a hundred years earlier, and look for the backyard patch where his unmarked grave still lies. But there was more still to come, because if Lincoln town was the birthday cake to end all birthday cakes, Fort Stanton was the icing on it.

Fifteen minutes up the road, a journey that took several hours back in the 1870's, another secret historical garden full of the most exotic treasures awaited. There were no tour guides, there was no entrance fee, no Keep Out or Don't Touch signs. Lonely, empty, Fort Stanton—and almost every stick and stone of it the real, original thing—was mine, all mine, to explore at will and to enjoy at leisure. Names flickered through the notebook of my mind: blustering Robert Widenmann, hot-tempered Jimmy Dolan, wily Pat Coghlan, devious Alex McSween, even the fiery Apache leader Victorio, they'd all trodden this same ground. And now, me.

Over there were the (surprisingly spacious) commanding officer's quarters which had once convivially sheltered characters as disparate as New Mexico Governor Lew Wallace and Col. Nathan A. M. Dudley, Kit Carson and Lawrence Gustave Murphy, Black Jack Pershing and (I do not doubt) the outlaw Jesse Evans. Across the neatly-mown "parade ground"– grass never grew there in those frontier times— was the long building where the infamous "Dudley Court of Inquiry" was probably held, not to mention the guardhouse in which, at one time or another, practically every fighting man of the Lincoln County War was incarcerated.

Further on, I found the old post hospital where many of the combatants in Lincoln County's various wars—the Horrell war, the Horsethief war, the Pecos war, the Tularosa ditch war, and of course the one that gave us Billy the Kid—were patched up, and where a few even died. Then, finally—hard to find and well off the beaten track—the original military cemetery with its scattered, mostly anonymous markers, each with a history and a thousand memories of its own. Bob Olinger and Bob Beckwith, both casualties of the Lincoln County War, are buried there, but no one knows where.

There was only one thing wrong with Fort Stanton in 1971: it was very slowly rotting away and nobody seemed to give much of a damn if it did. To be fair, sporadic attempts were made later to save it, notably one by the Lincoln County Heritage Trust in 1988, but there was never any concerted, planned and properly-directed action.

Well, then was then and now is now.

Bravely and boldly led by the author of this book, a determined, dedicated group of people who care deeply about its future has taken up the cause of Fort Stanton, and the difference they have made is astonishing. In 2004, Lynda took me on a tour of the site, more to show me the sheer size of the task than to convince me of the necessity of doing it. In 2007, she took me back to update me on the struggle; to see how much has already been achieved by dint of unstinting hard work and countless hours of their time freely given. The steps forward that have been taken are truly giant ones, but everyone involved knows that over the past half century the entire complex has been used and abused so badly that restoring it will cost millions. It sounds impossible, but they believe it can, it should, and it must be done.

And why?

Because there is no place anywhere more redolent of the whole history of the Southwest, of the lives of the men and women—military and civilian, white, Hispanic and native American—who peopled it from its establishment in 1855 through the Civil War, through the dark days of civil unrest and frontier gang warfare stretching from the Horrell War of 1873 to the death of Billy the Kid in 1881, from the bloody decades in which the Mescalero Apaches rebelled and visited hell upon the white settlers of New Mexico and Arizona, to the Fort's final abandonment by the military in 1896, its reincarnation as a Marine Hospital, a prisoner of war camp, an open prison ... well, enough. That there is a lot of history at Fort Stanton is not the issue. The issue is, does it matter?

I believe it does...

We have a very simple choice to make: we can save Fort Stanton or we can let it go. The latter would be an irreparable loss, but I do not doubt there will be those who argue for it. Nevertheless, I passionately believe it can—and it must—be preserved, because in many ways it is as important an historical site as Gran Quivira or the Alamo, Monticello or Mount Rushmore. By helping to establish the Fort Stanton State Monument, Lynda and her team have succeeded in erecting around it a partial mantle of protection. But it is a fragile

(Opposite page photo) A "Spartan" Lifestyle on the frontier as displayed at the Fort Stanton Museum.

Photo by David L. Tremblay

one, and it needs immediate and continuing support. Having spent more than fifty years of my life trying to establish some of the truth about the past in which Fort Stanton played such an important part, I am not ashamed to urge you, to implore you, to beg you to help. You have already begun by buying this book. Now go the extra mile. Write to the Governor of New Mexico. Call your Congressman. Petition your President. Tell them "fine words butter no parsnips." Tell them, in the name of your children and your children's' children, that this magnificent piece of American history must be preserved and restored properly. A country that turns its back on its past has nothing with which to face the future.

Photo by David L. Tremblay

Fred Nolan

Chalfont St. Giles, England

March 2, 2009

Frederick W. Nolan, perhaps the premier expert and enthusiastic supporter of all things pertaining to the Lincoln County War, has written several books about the subject. At a very young age Nolan wrote the biography of John Henry Tunstall, who also died a very young man in that conflict. This was but one of many intriguing steps back in time and Fred brought new insight into many arenas of this most violent of struggles, the infamous Lincoln County War. He is perhaps the leading British specialist in Western Americana yet he can write thrilling novels about everything from the Russian Revolution to the old west. Additionally, over the last four decades, Fred Nolan has become a friend to many of us in Lincoln County. We have come to know his lighter side as he participated in the McSween site Dig in 1987. He also played the piano and sang for us at various gatherings in old Lincoln town over the years. A true Renaissance man, Fred Nolan is our kind of man!

Courtesy Fred Nolan

Photo by David L. Tremblay

Photo by David L. Tremblay

Photo by David L. Tremblay

Sierra Blanca
Photo by David L. Tremblay

Chapter 1

Following Red Tailed Hawk

(A Brief History of Mogollon and Apache Peoples)

The haunting cry of Red Tailed Hawk hovering over lands dominated by the Sacred Mountain, Sierra Blanca, reminds us that the territory over which this magnificent bird flies today has changed little through the centuries. His bird's eye view would reveal an ancient land and a culturally rich New Mexico dependent upon its natural and cultural heritage. Encouraging others to follow along the trail of the Red Tailed Hawk and to learn about the native population that considered this bountiful yet harsh land their home long before the Spanish and Anglo pioneers arrived is an adventure in itself. The Mescalero Apache still live here today and are part of the colorful mosaic of peoples inhabiting the region. Fort Stanton and its community came much later, establishing a foothold in a land once ruled only by the Apache and Red Tailed Hawk...

According to the Mescalero tribal brochure, Sierra Blanca symbolizes "a place called White Mountain…where the creator gave us life and it is a special place. It was on White Mountain, according to legend, that White Painted Woman gave birth to two sons, Child of Water and Killer of Enemies….When they grew up to be men, they rose up and killed the monsters of the earth. There was peace and all human beings were saved." Since their creation at White Mountain, the Mescalero have gathered resources in mountain refuges allowing them to survive the hot and arid climate of the region.

Sierra Blanca also provided this same sacred ground for the Mogollon peoples who came before the Apache…

THE JORNADA MOGOLLON

Ceramic styles of the Mogollon: Mimbres (often used as trade ware), Lincoln black/red, Chupadero black/white.

Courtesy Tremblay & El Paso Museum of Archaeology.

The once dominant nomadic Desert Culture known as Clovis gave way to the more sedentary Mogollon, who resided in southern New Mexico, Arizona and northern Mexico more than one thousand years ago. The Mimbres branch of the Mogollon lived near Silver City and were famous for their anthropomorphic and stylized pottery. In fact, until recently, the Jornada Branch of the Mogollon was considered somewhat like the poor step child, the one not so gifted or artistic. However, along the Bonito, Hondo and Ruidoso Rivers of Lincoln County, small pit house villages and later larger pueblos were scattered throughout, and recent discoveries of large ollas (water jars) in beautifully designed Chupadera black/white or Lincoln black/red design motifs indicate a highly artistic factor no one ever suspected.

For centuries populations ebbed and flowed within this harsh land of abundant harvests and devastating drought. Because of this "feast or famine" situation, the Mogollon relied less on agricultural activities and more on hunting and trading, even as far as the Plains area of West Texas and northern Mexico. However, clashes with the Plains Indians and also some of the Meso-American tribes may have led to their abandonment of the region, especially once drought set in for more than a century.

Where did they migrate to after deciding life was no longer secure from enemies or from drought? The region was nearly abandoned by the late 1400's and archaeologists theorize many may have moved north along the Rio Grande, joining their northern neighbors in various Pueblos that also survived the influx of ancestral pueblo people (Anasazi) moving south to escape the devastating drought in their lands. Other, smaller groups may have drifted to northern Mexico.

Some claim that the Suma and Manso bands who greeted the Spanish explorers could have been descendants of the Jornada Mogollon. About this time the people known as Athabascans (today's descendants being the Apache and Navajo tribes) were making their way from Canada and Alaska into the southern reaches of the Southwest. It was a time of change and great turmoil.

In terms of religious features, beliefs and ceremonies, the Mogollon had their kivas, as did all Southwestern sedentary folks. However, along the Bonito River, Feather Cave** and Arrow Grotto among many other smaller cave locations, signify an incredible wealth of unique rituals outside the secrecy of the Kiva.

Hand prints, pictographs and petroglyphs have been discovered along rock outcroppings and canyon walls as well as within the depths of caves. Carved *pahos* (prayer sticks), miniature and full-size bows, arrows, sandals and other offerings contradict previous evidence that they were devoid of a rich and complex theology. Rare Macaw feathers and other trade goods and design motifs also indicate major contact with and influence of their Mexican neighbors.

Cupules (a form of rock art or special culturally utilized holes along rock faces or canyon walls) and solstice locations are also found throughout the region.

Three Rivers Petroglyph site on the other side of Sierra Blanca from the Bonito contains more than twenty two thousand petroglyphs including sun bursts, masks, hand prints, geometric designs, wildlife, reptiles, a man inside the belly of a huge fish, goggle eyed beings and horned figures. Many scholars believe these latter motifs validate a major Meso-American connection. Research into these magnificent petroglyphs is on going. Who knows what secrets will be revealed?

***Author's Note:* Lincoln State Monument now houses an exhibit displaying part of the items from Feather cave and Arrow Grotto.

Photos by David L. Tremblay

Photos by David L. Tremblay

THE APACHE PEOPLE

At about the same time that the sedentary culture of the Jornada Mogollon was disappearing, a wave of nomads made their way into the greater Southwest, eventually becoming a significant force of conflict within the Native American community and with the earliest European settlers. No history of Fort Stanton can be written without understanding the general history of the Apaches and why they fought so hard to keep their homelands from the invading White Man.

The Apaches had traditionally raided Pueblo villages for women and food supplies and also Mexican and American settlements for horses and ammunition and other goods they could not obtain from the Pueblos. They raided deep into Mexico and as far north as Colorado.

Photos by David L. Tremblay

Although there were many bands of Apaches scattered throughout the Southwest, those who dominated the lands within range of Fort Stanton were the Mescalero. Named by the Spanish for their gathering of the mescal plant and its use as a major food source, they became known as mescal gatherers or Mescalero.

Mogollon petroglyphs at the Three Rivers BLM Site in the Tularosa Basin.

They were divided into two groups. A Plains-like culture raided and lived in Texas, especially in the Fort Davis, Big Bend and Guadalupe Mountain areas. Their primary leaders were Juan Gómez, Peso and Natzili.

The other bands lived in the area known as Sierra Blanca or the Sacramento Mountains. Referred to as the "Sierra Blancas" to help differentiate their people from the more warlike Mescaleros further south, they sometimes joined forces or traded, but most of the time they went their independent ways. Their leaders were: Cadette, Santana and Sombrero, (son of Natzili) and Solon Sombrero, grandson of Natzili. This was Natzili's old band that had eventually intermarried with the Sierra Blancas, thus deciding to leave the southern reaches of their territory and stay within the shadows of Sierra Blanca.

From the arrival of the first Spanish Conquistadores, through the era of Spanish control, Mexican independence and Anglo pioneers flooding into their homeland, it is understandable that hostilities would emerge. Lives would be lost and many battles would be fought. From 1540 until 1886 the area was a maelstrom of conflict. Comanches also began to crowd the Mescalero and Lipans as each group vied for new territory. Despite the "presidio" system of Forts throughout their empire, Spanish and Mexican soldiers continued to fight and defend what they believed was their territory, won for "the glory of God and the King of Spain." They tried everything from extermination to reservations for the nomadic Apaches. Nothing seemed to work for long. The territory was a vast land from Coahuila and Chihuahua in Mexico to the Colorado borderlands, Big Bend country in Texas and the Guadalupe, Sacramento and Sierra Blanca ranges in southern New Mexico. There were never enough supplies or soldiers to defend such a vast, frontier land.

In 1848 New Mexico and much of the Southwest and California were ceded to the United States via the Treaty of Guadalupe Hidalgo. By the early 1850s Hispanic families, mostly sheepherders and small agriculturists, began moving into placitas or villages along the river valleys in Lincoln County. *Torreons* or two-story towers became defensive shelters in times of attacks. (An excellent example of a *torreon* can be seen in historic Lincoln.) Cultural clashes occurred, yet more settlers continued to arrive, thus providing a recipe for disaster.

Courtesy Lynda Sánchez collection.

The Four Warriors in field dress. Geronimo on far right. 1886.

1855 was a pivotal year for the war-ravaged land. Fort Stanton was finally established along the banks of the Rio Bonito (see Chapter II on Fort Stanton for specifics) to control marauding outlaws and the Mescaleros. Death, raiding of livestock and communities, and basic harassment continued, despite efforts by the military to settle things down. The selling of liquor to the Mescaleros did not help the situation and bootlegging became a common occurrence. Drinking was also a problem on military forts of the time, and enterprising locals got rich selling illegal liquor to both sides.

During the Civil War both Union and Confederate troops roamed the state and for a short time the Confederates controlled Ft. Stanton. However, after Kit Carson, fresh from victory at Val Verde, came marching into the area, the Confederates fled south. Furthermore, they were weary of clashes with the warlike Apache.

Shortly after his arrival, Carson received stunning orders from General James Carleton, who, in his absolute desire to destroy the Mescaleros, had virtually no mercy or concern. Ordering Carson, in an infamous and often quoted 1862 letter, to see that: *"All (Mescalero) men were to be killed whenever and wherever"* they could be found. Carleton also noted that women and children were not to be harmed, but to be taken prisoners and held at Fort Stanton until Carson received other instructions. Those instructions were not long in coming, for by December of 1862 Cadette, along with several other leaders, sued for peace, and the tragedy known as The Bosque Redondo was about to begin.

Luckily the Mescalero who knew Carson would not believe that he would murder their people. They decided to head to Fort Stanton and camped close by. Sometimes in today's view Carson was a villain, but we need to note that Carson just 'seemed to forget' those vicious orders from Carleton to kill the Mescalero "whenever and wherever". In other words those who made it to the fort were fed and protected by Carson and his troopers until the orders came. Eventually, thousands of Navajos and about 475 Mescaleros were forced to live together along the muddy and sluggish Pecos River near Fort Sumner.

Just prior to their forced march to Fort Sumner, Cadette made an impassioned speech and plea for his people to Carleton that was translated by Captain John Cremony...

> *"You are stronger than we. We have fought you so long as we had rifles and powder; but your weapons are better than ours. Give us weapons and turn us loose and we will fight you again; but we are worn out; we have no heart; we have no provisions, no means to live;...You have driven us from our last and best stronghold, and we have no more heart. Do with us as may seem good to you, but do not forget that we are men and braves." Carleton showed no compassion...*
>
> (Terrell, 235-236)

Courtesy Lynda Sánchez collection.

Mescalero camp. c. 1900. Many such camps surrounded Fort Stanton prior to the Bosque Redondo.

Cadette and the Mescalero sadly departed their homeland. However, there were about 100 who refused to give in and they "disappeared" into the rugged canyons or joined with Victorio and Geronimo.

Perhaps one of the best descriptions of that hell hole (Fort Sumner) came in an interview with Scout Big Mouth who was a small child at the time of this major event in the lives of the Mescalero:

> *"That place, Fort Sumner, was what is now called a concentration camp. There was nothing there for us except misery and hunger. There*

Photo 1938, by Dan Kusianovich

Scout Big Mouth.

were no pines, no streams…and no game. There was no water fit for drinking. We had been accustomed to the clear, cold water from the melting snow of Sierra Blanca…We had to drink the muddy, ill tasting water from the Pecos. It made us sick; it even made the horses sick. ..

"They put men and women to work digging ditches and digging up ground with shovels to plant corn. And once a week the soldiers gave us enough food to last perhaps two days. We were not farmers – we were fighters and hunters! Above all, we were free people; and now we were imprisoned within picket lines and made into slaves. The Apache does not mind work, but he does not like slavery. We were cold, hungry and miserable. Above all, we were homesick for our country and our freedom…"

(Ball, Henn and Sánchez 1980: 202)

It was a black chapter in the history of native peoples of New Mexico. Not only was the area around Ft. Sumner depressingly dry and isolated, hunting was not allowed, and the water made people sick. Disease was prevalent, especially the scourge of small pox. Apaches who survived remembered trying to obtain water for their families along the banks of the Pecos and seeing bloated bodies of pestilence-ridden Navajos floating on down to who knows where. That had been the last straw. Leaders calmly told their people to get ready, to pack quietly and to slip away, one group at a time. And thus began the trek home to the sacred lands around Sierra Blanca. Even Fort Stanton looked good compared to what they had been through on the Pecos.

By 1866, observers agreed this holding area was a disaster. Finally Carleton was relieved of his command. Before the military returned the Navajo to their traditional lands (1868), most of the Mescaleros had already disappeared into their mountain hideouts, vowing never again to be taken to such a terrible place. Shortly afterwards even military officials recognized their miserable failure and the Bosque Redondo was abandoned.

Geronimo. Always strong and defiant, his men were the last to surrender in 1886. Outnumbered 5000 to 17, Geronimo and his fighting men were beyond brave.

Eventually a reservation was established in 1873, giving the Mescalero the heart of their homeland, with Sierra Blanca, one of their sacred mountains, as part of this reserve. Fort Stanton was by now well established; however, conflict continued and became a way of life measured in sporadic raids on villages and retaliatory raids by white farmers and the military. There was never enough food and other supplies were severely rationed. Many Mescaleros stayed hidden for years in the mountains rather than come into the newly formed reservation. They were blamed for every raid or crime committed, or so it would appear, yet the White Man was rarely if ever punished for stealing their horses, raiding their ***rancherias*** or for selling bootleg whiskey.

On the periphery were valiant warriors like Victorio, Geronimo and others who were suspicious of anything promised by the White Eyes. A few, like Dr. Joseph Blazer and family, were trusted. But they were the exceptions. Blazer's Mill (La Maquina) was a major landmark and source of labor and supplies. This family, from the beginning, had integrated into the daily lives of the Mescalero. Their descendants today play major roles in tribal policy.

Victorio, a Warm Springs Apache, often came into the reservation lands despite Fort Stanton and the military being so close. He came with his people for rest, for supplies and perhaps with some hope of being able to settle down and live as they once had. But that was an ill-fated dream for the followers of Victorio. Before the final flight from Mescalero, Victorio shook hands with Dr. Blazer and noted with regret that they would never again be confined to the boundaries created for Apaches by

Mrs. Paul Blazer and her pupils, Bent, New Mexico.

The Blazers worked with Santana to build bridges between cultures.

Courtesy Lynda Sánchez collection.

White Men… *"From now on it will be war, war to the death. There is no other way."*

During this time Santana, a war chief of the Mescaleros, was a strong leader known to the Blazer family. For years he was unknown to the military although he was a contemporary of Mangas Coloradas. According to some, he was one of the more notorious of the Mescalero leaders, responsible for many of their most vicious crimes. He was for innumerable years a "shadow" figure, an unknown who held out against the white man and he was definitely one of their most implacable enemies! Yet, in the end, the friendship and integrity of Dr. Joseph Blazer and Santana's insistence upon good land for his people helped to create a safe haven for them. We see no official mention of him, however, until the late 1860s.

Dr. Blazer's family story also parallels Santana's as their two worlds merged. Neither had learned the language of the other; however, a powerful confidence and cooperation were forged as they struggled to communicate in English, Spanish and Mescalero. Along with many acts of kindness, trust eventually developed into a binding friendship, and Santana and his people came out of the shadows.

Unfortunately, more bridges needed to be created but during those times, as is so often the situation, *"the peaceful and innocent suffered for the actions of the more warlike Mescaleros. General (Edward) Hatch ordered the entire Mescalero tribe disarmed and sent troops to the agency for that purpose. They killed many including ...one small helpless group near the Mescalero agency who were mercilessly slaughtered."* (Terrell, 1974, page 340) This cycle of violence confused the Sierra Blanca Mescalero, so they remained hidden in the mountains. It was, no doubt a wise decision for their survival.

Courtesy Lynda Sánchez collection.

James Kaywaykla

Victorio and his followers, including Mescaleros who chose to raid with him, continued to fight running battles from the reservation territory, through the Black Range and San Andres, creating a trail of bloody devastation. Finally, crossing into Mexico in October of 1880 they were either killed in battle as was appropriate for a warrior, or captured at the Battle of Tres Castillos in northern Chihuahua. The captured ones were marched to the city of Chihuahua where those who survived were sold into slavery. Victorio was not among them.

Another of the more intriguing characters of the Apache Wars was James Kaywaykla, who, like Victorio, was a Warm Springs Apache. However, his life was intertwined with the Mescaleros, including relatives, friends and life styles, especially after the Mescaleros accepted the last of the Chiricahua POW's as residents of their reservation.

During many interviews with author/historian Eve Ball, he stressed that: *"Until I was about ten years old I did not know that people died except by violence. That is because I am an Apache...whose first vivid memories are of being driven from our homeland".* (Ball 1970: xiii)

Kaywaykla passed on to the "Happy Place" of the Apache in 1963, but not before he helped Eve Ball garner a better understanding of what life was like for his family and his people. As a child he had experienced an early life of violence, betrayal and tragedy. He survived, however, and for that we must all be grateful for we now have much of his story, his side of events in written form that can be passed on to those who have an interest in our regional history.

Much later, after the Apache Wars ended, (in 1886) the Mescalero generously opened their reservation to the Lipan and later to the Chiricahua Apache POW's who needed a place to come back to after their 27 years of imprisonment (1886-1913). They had suffered and died in Florida and Alabama. Fort Sill, Oklahoma was somewhat better but could never compare to their beautiful mountain homelands... During this time of change, another strong character emerged. He was a white man who bore the burden of righting many wrongs. A gentle giant of a man, revered by both sides, by young and old, Apache, Hispanic and Anglo, Father Albert Braun, long an advocate for the Apaches, came to Mescalero as a very young priest in 1916. He, too, met many of the survivors of the Apache Wars and throughout his long life that almost reached the century mark; he began to learn their side of history. He concluded in his foreword to ***In the Days of Victorio:***

> *"Much has been written by the White Man about the Apache... accounts... written as military reports by young officers Ambitious for promotion; reports were... compiled by Indian agents; by contemporary newspapers whose owners depended upon advertising paid for by merchants who lived by selling supplies to the reservations, by pioneers who entered the country occupied and claimed by the Apaches... Some were written by men who wanted to take over Apache country... seeking the support of public opinion, these intruders often picture the Apaches... as a people who should be driven from the mountains of the Southwest."* (Ball, 1970: vii and viii)

These same people depicted the warrior folk as vicious, cruel beyond measure and totally uncivilized.

"Partial truths" are really lies and soon Father Albert and numerous others began to flesh out the entire culture and history of the Apache people. There were enough of the tribal elders who described the early days and who entrusted some white people with their folklore, religious beliefs, and their side of events. Archaeological records and other ethnographic accounts fill in additional pieces of the puzzle.

World War I and II created numerous changes on the reservation as elsewhere in the United States. Educational opportunities, off-reservation employment, boarding schools and utilization of their tribal resources also created change. Some wounds from the Apache Wars were beginning to heal, and understanding of cultural differences began to occur.

Courtesy Lynda Sánchez Collection.

Ramona Chihuahua Daklugie in her wedding dress.

The Mescaleros have created a tribal historic preservation office whose responsibility is "to maintain and research ties with the past". Professional archaeologists are on staff and a tribal cultural center exhibits many intriguing artifacts from their past.

During the Fourth of July they open their reservation to Rodeo, a popular sport among many tribes. They also allow the public to view their puberty ceremonials and other celebrations. The puberty ceremony is one of the most vivid and beautiful coming of age celebrations anywhere in the world. The young girls represent White Painted Woman, mother of Child of the Water and Killer of Enemies. She provides the moral compass for the girls. Family members participate in preparations from sewing buckskin dresses, cooking huge amounts of food, purchasing gifts, and other family oriented projects for many months prior to the actual ceremonial. It lasts for four days, and is usually open to the public for most of that time. Very little has changed from the past centuries.

Today the tribe owns a ski resort, a casino and the Inn of the Mountain Gods. They have cattle and timber resources and one of the best forest fire fighting units anywhere - the Red Hats. The reservation lands occupy over 475,000 acres and there are more than 5,000 tribal members in residence, most of them on the reservation. Thus, as the beginning years of the 21st century are upon us, we observe a vibrant tribe, the Mescalero, who are prosperous and eager to face the future on their own lands and on their own terms with their sovereign government in place with talented and powerful leaders. Wise tribal elders understand that their traditions must continue, their language must be maintained, and that pride of culture can only continue with a strong corps of adept and skillful individuals who will lead their people into the future.

Mescalero Mountain Spirit Dancers.
Fourth of July 2008, continuing in tradition.

Photo by David L. Tremblay

Building a bridge between and among cultures is easier now despite the fact that many of the Apache still feel animosity and anger when any Fort is discussed and especially Fort Stanton or Fort Sumner. However, Father Albert's vision when he returned from France after World War I, and later from World War II where he had survived Corregidor and had been incarcerated as a POW along with the men of the dreaded Bataan Death March, was to create a mission to serve his people at Mescalero and to build some of those bridges he knew were important to heal the wounds of many wars and cultural misunderstanding. Through years of blood, sweat and many tears his vision came to fruition. From humble beginnings after WWI, to the now grand monument dedicated to all Veterans of both wars, in 1945 it was referred to as an "impossible dream translated into soaring stone". Anyone who visits or who has been involved with the ongoing restoration work at the Mission today acknowledges that it changes their lives forever.

Illustration by Theron M. Trombeau.

Saint Joseph's Apache Mission and Veteran's Memorial.

The restoration crew has worked for several years not just to restore this magnificent church, stone by stone, but to instruct young people in a trade that builds self-esteem and knowledge of their heritage. The project provides a cornerstone and a bridge to cultural gaps as both Mescaleros and their Anglo and Hispanic neighbors work side by side in this preservation effort. The restoration team and all participants are reaching out to others including those of us working to preserve and restore Fort Stanton. We are proud to work side by side with these incredibly dedicated individuals like Mary Serna and Tommy Spottedbird. We have, by working together, discovered that all of us have, indeed, a common battle to preserve our past. Dr. Blazer and Santana would be pleased!

And, Father Albert would certainly agree that by preserving this Veteran's Memorial and national landmark, the Mescaleros are working through a *"process ... that strives to honor their ancestors, empower their contemporaries and inspire their children."*

Courtesy Saint Joseph's Restoration Project.

Courtesy Saint Joseph's Restoration Project.

Above: Father Albert Braun, Chaplain WWI and WWII. Laid to rest inside his beloved Mission.

Left: Tommy Spottedbird, Restoration Co-Director and Job Foreman.

Opposite: Saint Joseph's Apache Mission Restoration Crew. (Left to Right) Mary Serna, Arylis Chee, Tommy Spottedbird, Brother David Paz, OFM & Brother Peter Boegel, OFM. This devoted crew has worked side by side with others over the past several years to lovingly restore this magnificent mission. They are an inspiration to all and have shown many of the Fort Stanton advocates that not only can bridges be built among peoples of all cultures, but that the Mission, like the Fort, is worthy of their time, labor and dedication.

Below: Mary Serna, Restoration Co-Director/Administrator and the "one who does everything else that needs to be done".

Courtesy Saint Joseph's Restoration Project.

Building bridges between and among peoples. This Veteran's Memorial represents "an impossible dream translated into soaring stone."

Courtesy Saint Joseph's Restoration Project.

Mescalero Apache Baskets.

"Baskets are the Indian woman's poems... They wove them into the story of their life and love." — Almer N. Blazer

Sources and Recommended Reading:

Interviews—

Eve Ball, many interviews, 1978-1984

Interviews by Eve Ball and mostly recorded in her books listed below (Percy Big Mouth, Scout Big Mouth, James Kaywaykla, Silas Cochise, Amelia Naiche, and dozens more.)

Books—

Ball, Eve, Henn, Nora and Lynda A. Sánchez, *Indeh, An Apache Odyssey*, Norman, University of Oklahoma Press, 1988

Ball, Eve, *In the Days of Victorio, Tucson*, University of Arizona Press, 1970

Blazer, Almer N., *Santana, War Chief of the Mescalero Apache*, *Ranchos de Taos*, Dog Soldier Press, 1999

Haley, James L., *Apaches: A History and Culture Portrait*, Norman, University of Oklahoma Press, 1997

Terrell, John Upton, *Apache Chronicle*, New York, Thomas Y. Crowell Company, 1974

Thrapp, Dan L., *The Conquest of Apacheria*, Norman, University of Oklahoma Press, 1967

Utley, Robert M., *Frontiersmen in Blue*: The United States Army and the Indians, 1848-1865, New York, The McMillan Company, 1967

Author's note: There are an incredible number of sources and books about Apaches. In addition to these sources, I would recommend checking out the bibliography in each of these works or just *Google* the topic.

Courtesy Lynda Sánchez Collection.

Mescalero Agency. c 1920.

Courtesy Lynda Sánchez Collection.

Mescalero campsite. Note teepee's. c 1920.

Courtesy Lynda Sánchez Collection.

Old Blazer's Mill. c 1920.

U.S. Military Forts: New Mexico, Southern Colorado, and West Texas

Photo by David L. Tremblay

Chapter 2

The Cavalry Era, 1855-1896

(Conflict and Challenge on the New Mexico Frontier)

While the sacred mountain, Sierra Blanca, dominated the physical and spiritual landscape of Lincoln County and provided the life blood of the region with its magnificent watershed, Fort Stanton proved to be the heart of the county. By providing economic opportunities, social activities, and defense for area residents, the Cavalry and the fort became their life support for decades..

"It was a fine sight, all with sabers drawn…It gives you a thrill and my eyes filled with tears…it is the call of one's ancestors…It seems to me that at the head of a regiment of cavalry anything would be possible…"

General George Patton, (D'Este, p 155)

Photo by David L. Tremblay

2005 Fort Stanton Live Celebration on the Quadrangle.

A grand 150th birthday celebration was inaugurated on the grounds of this incredible site in 2005. Visitors were allowed to step back in time for a brief sampling of life during the 1800's. Men and women dressed in period attire paraded about, and the folks came from near and far to admire, to learn, and to enjoy historic exhibits, Apache dancers, reenactors from both Confederate and Union sides, Buffalo Soldiers as well as music and dancing from that bygone era. Innumerable dedicated volunteers worked to make this a very special occasion. They also were inspired by a dream in the near future when this crown jewel of military forts will be preserved and open to the public for many generations to come.

The mists of time have long since erased the thunder clouds of Civil War, of range wars and of renegades rampaging through the Bonito Valley during the violence and cruelty of the infamous Lincoln County War and Apache conflicts. The Fort was a strategic location held by both Confederate and Bluecoat. Valiant Apache warriors, Buffalo Soldiers, Hispanic and Anglo sheepherders, ranchers, merchants and outlaws are all forever etched into the collective memories of area residents. Their history is tied to this fort along el Rio Bonito and both locals and visitors acknowledge that sometimes they can almost hear the bugles and marching troopers as they ride into the distance.

Courtesy Lynda Sánchez Collection.

Rafael Chacon (first Hispanic Commandant at the Fort).

If Rafael Chacón, Kit Carson, Black Jack Pershing, or even Billy the Kid and the McSweens or Governor Lew Wallace, author of the famed "Ben Hur" were to return today, the setting would be about the same. Framed by towering Sierra Blanca and the Capitans and meadows of blue gramma grass, they would recognize the elegant Victorian porches, some with columns, carved doors and window frames and a parade ground to honor her soldiers. They would recognize the great care and craftsmanship in construction features, of fireplaces, marble top mahogany furnishings, of beautiful wooden staircases and tin ceilings. The stables made of sturdy stone walls, still stand after more than a century of use.

The quadrangle contains buildings representing other eras as well. Pueblo Revival architectural styles, Depression era bungalows and even a small chapel of native stone stand proud as evidence of a rural community struggling to make things comfortable and pleasing to its residents.

Photo by David L. Tremblay

Re-enactors at Fort Stanton Live.

Photo by David L. Tremblay.

Fire Power.

(above) Kenneth Dusenberry, Artillery Company of New Mexico

Dover Collection.

Buffalo Soldiers in camp. Frederick Remington.

There are many phases to this Fort's history, each one contributing to the unique heritage of New Mexico and the early pioneers. Without its base of strength and need for produce, hay, corn and basic supplies, the region would have declined economically long before tourists arrived to marvel at the awesome landscapes, night skies and historic sites.

Fort Stanton stables today.

Photo by David L. Tremblay

MILITARY BEGINNINGS (1855-1896)

Fort Stanton had a distinguished and, for a military fort, a long history that continues to the present. First established in 1855 as protection for settlers from Apaches and marauding outlaws, it was built of native stone instead of the usual adobe construction of New Mexico forts.

Courtesy Lynda Sánchez Collection.

General Black Jack Pershing.

Named for Captain Henry Stanton, who lost his life during a battle with Apaches, the fort provided a major market and added to the prosperity of the region. 1860 census figures listed 257 residents along the Bonito, most being of Hispanic origin. There were small ranches and farms and a variety of occupations was listed including one musician, two teamsters, two carpenters, two millers, a black smith, and many herders.

Building the fort was no easy assignment and there were shortages of tools, work animals, tradesmen and supplies. Poor roads were another problem. Adobe was to have been the original material; however, the adobe was too brittle and it was decided to turn the adobe "layers" into stone "quarrires". Months of frustration finally saw the quadrangle completed. Roofs were of wooden shingles and most floors were of hardened earth. Hundreds of trees were planted over the years to soften the surrounding environment and to provide shade from the scorching New Mexico sun.

Life was definitely a struggle for this isolated frontier post, yet it was a much better locale than the majority of Forts in the territory. Constant skirmishes, raids and counter raids against the Mescaleros and outlaws kept the men busy and on guard.

The few women who came to live there did not last long, although Lydia Spencer Lane, wife of Lieutenant William Lane, described their duty at Fort Stanton as follows:

> *"Fort Stanton was a beautiful post, with the best quarters in the army at that time, but it was like being buried alive to stay there. Nothing ever passed that way, and it was seldom a stranger came upon us. There was but one mail a month. The Mescalero Apaches were in camp that winter near the post, and came and went as they pleased, walking into our houses and sitting on our porches without the least hesitation...I never could become adjusted to the Indians staring at me through the window when I was sewing or reading."*
>
> This description was penned in 1858.

The Civil War brought more problems, abandonment and destruction. However, that was short lived and once the Union regained control, the Fort was rebuilt. In 1864 the commander hired civilians to repair the fort, allowing the troops freedom for military operations. This move also pumped money into the local economy. After the Bosque Redondo fiasco the District of New Mexico agreed to allocate more than $20,000 for the rebuilding of the post. Money was for repair of the old stone structures and the building of new ones. LG Murphy was given a contract to supply 200,000 board feet of lumber. Henry Harrison was to supply 500,000 wooden shingles. Renovation attracted even more civilian workers and to this day Fort Stanton provides many jobs contributing to the economy of the region.

The military years from 1855-1896 were tough for the soldiers and especially so at Christmas because most of the men did not have their families with them. If there was a family on base or men had wives with them, caroling and parties at the officers' wing or perhaps at a local ranch house were part of the celebration. One New Years' party thrown at the Fort lasted until three in the morning! Perhaps having three fiddlers playing as people danced the night away might have been the reason. Sometimes regimental bands toured the frontier forts and several came through Fort Stanton greatly improving morale.

A tuba on display at the Fort Stanton Museum - 2008.

Lincoln, a short ride away, played an integral part during the holidays for the troops. Midnight Mass was celebrated and many of the Hispanic homes welcomed soldiers and others during the holidays. *Bailes* (dances) were part of the season and everyone had a grand time. A generous meal of red chile, tamales, empanadas and enchiladas was served. One can imagine how much the lonely men at Fort Stanton appreciated the generosity and the time to socialize.

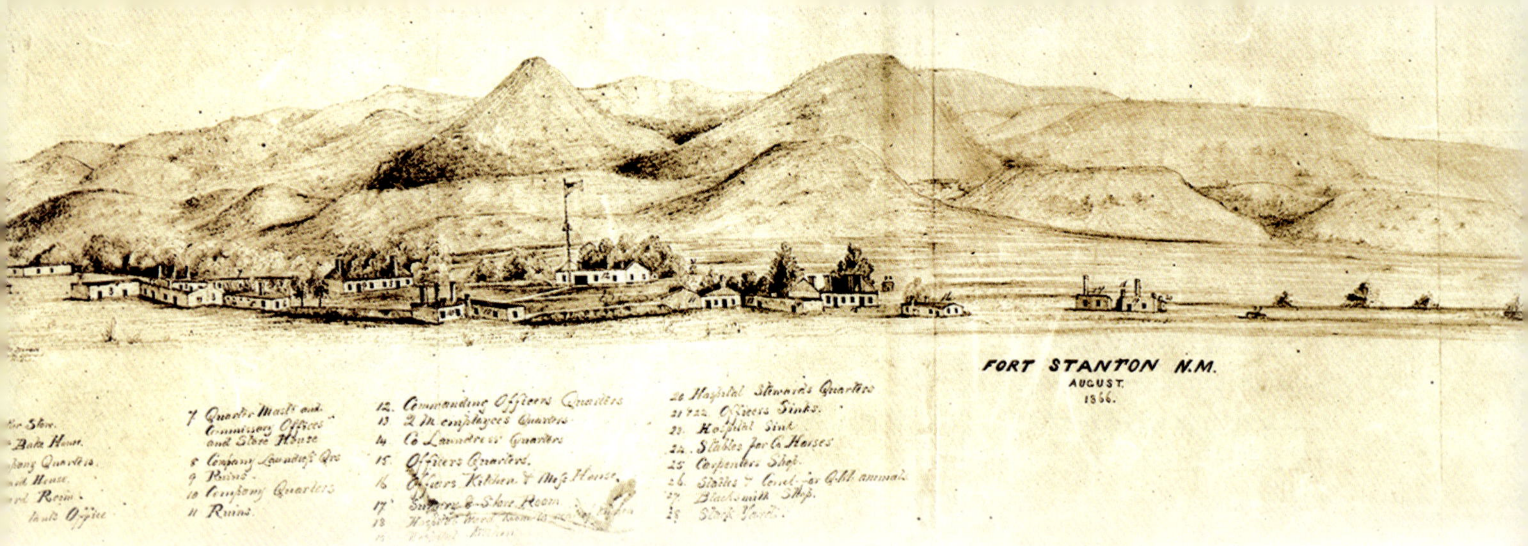

DUEL AT FORT STANTON

(The Paddy Graydon Affair)

(Above) 1866 drawing of Fort Stanton by Arthur W. Moore, 1840 -1913. He served as "hospital steward", a medical student and was originally from England. Later returning to Rochester, N.Y. he became a promient artist and supporter of the arts.

Army life, the good, bad and the raw, ugly conflict of the military and the stress the men experienced were exemplified by the so called ***Paddy Graydon Affair***. Good men with differing viewpoints on how to handle the Apache situation and the loyalty some men engendered are all part of the tragedy. Furthermore, this particular event was chosen to assist the reader in understanding the kinds of men that were present on the frontier during this time. They were tough, unapologetic and sometimes wrong, but they made a difference for the future and as with many controversial figures, one has to judge them within the context or time frame of history, not how we view things today.

Despite many people loathing Maj. James Carleton, when he assumed command of the New Mexico territory, he knew that once the Confederate threat was minimized he had to re-establish, and quickly, command of the Forts that had been abandoned, thus allowing his men to handle the growing problems with the Apache and Navajo. He wrote in September of 1862,

> *"...to punish and control the Mescalero, I have ordered Fort Stanton to be re-occupied. That post is in the heart of their country, and hitherto when troops occupied it, those Indians were at peace. I have sent Colonel Christopher Carson with five companies of his regiment of New Mexico volunteers to Fort Stanton."*

Under escort, Company K was to bring supplies needed to rebuild the Fort and that escort was lead by one Captain James (Paddy) Graydon and his Independent Spy Company. They were smug in the knowledge that the Mescalero had helped create havoc among the Confederates who were totally unprepared for their type of skirmish tactics.

(Fort Stanton: An American Journey, courtesy the Hubbard Museum of the American West January, 2009)

Paddy Graydon was considered to be one of the real characters of the Civil War scenario in New Mexico. Coming to the United States as a 30 year old in 1853 he, like many Irishmen forced to flee the starved out nation of Ireland's potato famine, enlisted in the 1st U.S. Dragoons. Heading west he had many adventures and clashes with Apaches in Arizona where he obtained an honorable discharge. After the Civil War began he re-enlisted only this time in the newly organized New Mexico Volunteers. His experience with frontier methods of fighting was welcomed by his commanders. Having routed the Confederates, he was now on his way to take care of the Apache problem at Fort Stanton. Little did he realize that this cocky attitude would have retribution.

Graydon had earned his reputation as being hard nosed, controversial, and colorful. Service with Graydon was a tough, no nonsense scenario with Indians, deserters, Confederates and outlaws. Because he was one of the few *"gringos"* who could speak Spanish, he got along well with the *Nuevo Mejicanos.* Although Rafael Chacon called him a braggart and arrogant he had recruited a good group of men from Lemitar and the general Socorro area. Soon these raw recruits were mustered into the federal army for a grand sum of 40 cents a day and a mount, if some of the poor creatures they had could be called that.

Marching toward Fort Stanton by October of 1862 they began to encounter Apache sign and eventually located a band of Mescalero under Manuelito near Carrizozo/Corona in the Gallinas mountains according to most reports. Some state the incident occurred at Gallinas Springs (Cement Springs). Supposedly the Mescalero demanded liquor and became angry and resentful when they did not get their way according to Graydon and several witnesses.

Other accounts claim they gave the Mescaleros booze and then shot them. Graydon was known to have little sympathy for Apaches. The one shot he fired unfortunately happened to kill their leader.

> In a 1923 article in the *CAPITAN MOUNTAINEER*, Judge George Kimbrell of Picacho was interviewed about his memory of the incident. He had entered the army *"a mere youth… had just come out to New Mexico from his home in Madison County, Arkansas, and entered the army under Carson. He was assigned to duty as a scout for Carson and when Carson started to move south he sent Captain Graden (Graydon) with a small force in advance of his command, among them, Scout Kimbrell, to look out for Indians. A small party of Mescalero Apache Indians was found in the Gallina (s) Mountains, now in this county. A meeting between Captain Graden and Chief Miguel Oleta (Manuelito) was held and another was arranged to take place the following day to try to arrange a treaty.*
>
> *"At the second meeting Captain Graden had sixteen men with him and Chief Oleta had eighteen. Captain Graden had instructed his men for each one to take an Indian aside under pretense of having a talk and he took the Chief. The Captain was armed with a double barrel shot gun and his revolvers; at a signal each man was to shoot his Indian, which was done and 16 were killed, the other two making their escape.*

Fort Stanton Archives

Kit Carson

> *"Captain Graden had learned through Scout Kimbrell that the whole tribe of the Mescalero Apaches, including their women and children, were in camp at Hondo Junction, where the town of Hondo stands today, and he wished to reach there before the two Apaches who had made their escape got to the main body and a rapid march was made,… when Captain Graden got in sight of the junction the Indians were moving out south. He pursued them through the Guadalupe Mountains, through southeastern*

New Mexico, killing several on the way, and through southwestern Texas. He headed them off from their route and harassed them back to Fort Stanton, where a treaty was made between them and Kit Carson." (CAPITAN MOUNTAINEER, December, 1923)

This was not the way to try and settle the "Indian" problem and create a more peaceful atmosphere. To make things worse, when they arrived at Fort Stanton Graydon's men found the fort in ruins since the locals had taken most of the windows, doors and any equipment that might have been left behind.

Courtesy Rio Grande Historical Collections, NMSU Library

Early Fort Stanton. 1871-1872.

However, this daunting task did not deter his men, and with his Company K and the five companies of New Mexico Volunteers they began to rebuild the fort. Much of the older parts of today's Fort were either built or rebuilt in 1863 once trade was established with the nearby communities. Many of the soldiers had carpentry and trade skills and he hired local individuals who had settled the nearby settlements. From October, 1862 until January of 1863 these men rebuilt the Fort to a place of real beauty. Construction continued by men of the regular army and later the Buffalo Soldiers.

Tragically all of this came to an end for Paddy Graydon as news got out about Manuelito's untimely death. On November 1, 1862 yet another major player in this affair arrived at Fort Stanton to visit his friend, Kit Carson. He was Army Surgeon John Whitlock. He had a good reputation and had been one of the doctors who saved innumerable lives of badly wounded soldiers after the Battle of Valverde. Whitlock had also married a local Las Vegas Hispanic woman and was the first Anglo to practice medicine in that community. He was originally from Kentucky and arrived in New Mexico during the 1840's.

Courtesy Bob McCubbin.

Post Sutler's Trader's Store – c 1880's.

Fort Stables – c 1880 – note some Buffalo Soldiers and others "grooming" their remuda.

Courtesy Bob McCubbin.

Post Sutler's Trader's Store – c 1880's.

Courtesy Rio Grande Historical Collections, NMSU Library

We continue here with Judge Kimbrell's version of events...

"Doctor Whitlock, the Regimental Surgeon and several others, including young Kimbrell, were in the Sutler's tent engaged at a game of cards. Dr. Whitlock there made some very severe criticisms concerning Captain Graden's conduct in killing the sixteen Indians at the meeting in the Gallina Mountains. One of Captain Graden's men was sitting in the tent at the time. He said nothing, but walked out and went directly to the Captain's tent and told him what the Doctor had said. A few minutes afterwards a paper was handed to Dr. Whitlock and on reading its contents, he wrote on the bottom of the paper, "Accepted."

"He then remarked, 'That is a challenge—I'll have to fight,' or words to that effect.

"Some of the parties present tried to get him to pass out the back way and go to Carson's tent, which he positively refused to do. He picked up a Colt's five shooting revolver lying on the card table, examined it, and about that time, Captain Graden was seen coming with a revolver in each hand. A meeting was arranged and the duel immediately took place, the running gears of a wagon standing between them, "Dr. Whitlock used the five shooting Colts and Captain Graden a Navy Six. Captain Graden had gotten his revolvers wet while (chasing the) Indians and only got it to fire twice, slightly wounding the Doctor. The Doctor then shot the Captain a dead shot. Just at that time, Lieutenant Morris of Captain Graden's company and about 75 soldiers rushed to the scene and commenced firing at Dr. Whitlock, who was crawling along a ditch trying to reach his tent, but he was killed before he reached it. He was hit by more than a hundred shots and after was dead and they threw stones on his body".

Understandably, Graydon's men had come to their commander's aid and carried the badly wounded man to a tent. The traditional account details more shots being fired and Whitlock fell, dead, having been shot by cowards, in the back, many times. Graydon's men threw Whitlock into what was most likely the irrigation ditch and others fired into his body. Carson later estimated dozens of shots were fired, perhaps over 100! This cowardly act was written up but nothing ever came of it. Three days later Graydon died and was buried up on the hill in the original cavalry cemetery for the Fort. He was later interred at the National Cemetery in Santa Fe.

Kimbrell's account notes that *"Great excitement was created and Carson ordered Lieutenant Morris and the others who participated in the killing of Dr. Whitlock arrested and had them disarmed and marched out on the parade ground and ordered them shot. Several of the officers present told him that would not do; that he would be court marshaled for such a course.*

"He (Carson) replied, 'Let 'em court marshal and be damned.' However, he finally yielded to their advice and revoked his order. Liet. Morris and one private soldier were court marshaled; the private was acquitted but Liet. Morris was found guilty and sentenced to be shot.

The night before he was to be executed, he made his escape from prison and was never seen afterwards".

Photo by David L. Tremblay

Old Cavalry Cemetary. Captain James Graydon buried here and later reinterred in Santa Fe.

Author's note: Judge Kimbrell, who was an eye witness to this "duel", provides by way of his excellent memory information that parallels the general historical account but adds intriguing bits and pieces to the puzzle that would have otherwise been lost. Both versions have been included herein to give the reader a feel for the trials and tribulations of the men who served. We would acknowledge that they were under great stress then, as even in today's military. George Kimbrell went on to marry a local Hispanic woman and had a huge family of his own. He became sheriff and later a judge. At present many still reside in Picacho, New Mexico.

Once things got back to normal after this tragedy, military life continued. Patrols were still needed to chase outlaws, rustlers and Apaches. The locals continued to depend on the Fort as their economic life line. They sold, traded and interacted socially with the men and women at the Fort. Furthermore, the role of the Fort became even more dominant as it was now the economic and defensive heart of the region, growing in importance each day.

"When we saw the smoke and heard the thunder And the rain of shells that fell on the gunners, We knew the Buffalo Soldiers were on their way With the light of day..."

(from Buffalo Soldiers by Christopher Mead, 1993, Fort Huachuca) one of many songs and stories about these brave fighting men.) At any given time only about 11-12 percent of the men at the Fort were Buffalo Soldiers. They did not come into NM until after the Civil War.

Courtesy BLM

Buffalo Soldiers along the Rio Grande.

Water from the Rio Bonito which flowed nearby was diverted to the local farmers' fields and to the Fort as well. According to author John Ryan, a report in 1870 describes water entering *"the fort by way of an aqueduct three quarters of a mile long, which ran to the southwest corner of the parade ground. From there an acequia, or ditch, surrounded the parade ground about twenty feet in front of the barracks and officers' quarters. ... the officers planted trees around the parade ground and dug the acequia to irrigate the trees. Upon exiting the acequia the water passed by the hospital via a single aqueduct and exited the post by way of the corrals, flowing back into the river again."* (Ryan, page 20)

Courtesy artist David Laughlin.

Buffalo Soldiers, 10th Regiment military drill.

Corruption and the Bloody Lincoln County War

Courtesy Lynda Sánchez Collection.

Above: Gov. Lew Wallace, author of Ben Hur.

Below: The Brewery - c 1880's.

The Fort was also a major source of conflict, and corruption during certain phases of its history. The wrong men in charge became greedy, and beef contracts and other supplies were fought over to empower them in their march to control the entire territory, especially after the establishment of the Mescalero reservation, and the culmination of the Lincoln County War. For example, by the summer of 1873, Murphy and Company was expelled from Fort Stanton *"because of their shady business practices with the Mescaleros and overcharging the government for rations supplied..."* A Captain James F. Randlett wrote, ... *"I consider that L. G. Murphy and Co.'s store is nothing more or less than a den of infamy, and recommend the removal of this firm from this reservation."*

Courtesy Lynda Sánchez Collection.

Upon arriving in New Mexico as its territorial Governor, Lew Wallace (October, 1877) had to deal with not only the rough and tumble life of the frontier, but the Lincoln County War/ Billy the Kid problems were literally thrown into the new Governor's lap. If it had not been for the distraction he had from his **Ben Hur** manuscript and then the arrival, one year

Courtesy Fort Stanton Museum.

later, of his wife he might have gone crazy himself!

Famed author/historian Marc Simmons, in his latest publication entitled STALKING BILLY THE KID, (2006) describes some of these events as follows:

"A year after taking up his duties, Governor Wallace, accompanied by Susan set out.... 'to investigate difficulties in remote regions of the territory.'....For safety on their journey, Lew and Susan rode in an army ambulance surrounded by a strong guard.....The Wallaces soon left the Rio Grande and veered southeast to Fort Stanton in Lincoln County. There, Susan Wallace wrote a detailed and interesting letter to their son back in Indiana. The opening lines of the letter, owing to her harsh criticism of the territory, have become famous.

Above: Classic Marching Troops on the Parade Grounds. c 1880's.

Below: Ruins of the Brewery.

Photo by David L. Tremblay

'General William Sherman was right. We should have another war with Mexico to make her take back New Mexico. I did not believe anything could make me think well of Santa Fe, but this hideous spot, Fort Stanton, does.'"

They were hosted by the Commandant and roomed in the quarters of the post trader. *'Efforts by Susan's hosts to please her were to no avail. They sent miles away for a chicken,'* she wrote. *'...I could not eat the skinny thing—all bone and fiber. I remarked to the post commandant that this country was not made for civilized man. He agreed.'*

She describes the outlaws, and other desperados, including Billy and his loud boasts about killing many men, and how he threatens the Governor himself. They kept their windows shuttered at night because the Governor wrote late into the evening hours. *'Billy has a gang of admirers and followers and they dash up and shoot out the candles... This way of living does not suit me. Sweet home in Indiana was never sweeter to my thought than now in this wilderness.'*

Finally she closes her list of complaints by stating, *'When the Ben Hur manuscript is ready we will go to New York with it. How glad I shall be to think there will be no return to this territory.'"*

Being a prominent military man (serving in both the Mexican War and Civil War) people felt he could deal with the outlaws and use the military at Fort Stanton to his advantage. Armed with a proclamation from the President he was, however, still loathe to push the martial law situation and only did so as a last resort.

A Riding Billy, by Walter Henn

Personal animosity with the commander of the Fort, Lt. Colonel Nathan Dudley, unfortunately created innumerable problems for Governor Wallace until he was finally forced to spend a considerable amount of time in Lincoln and at the Fort. His pardon of Billy the Kid and that fiasco has been written and discussed in dozens of books but the bottom line is that Governor Wallace finally ended the major feud known as the Lincoln County

War although the scars still exist today among some of the old timers.

Dudley was transferred to Fort Union, and Wallace breathed a sigh of relief even though martial law and "ex-parte" arrests by the military continued. An unknown correspondent sent a dispatch from the Fort to the Mesilla Independent stating:

> ... *"the guardhouse here is a 'Bastille' crowded with civil prisoners...both officers and enlisted men dislike this thankless duty."* (**Fulton, 342**)

Eventually, the gunsmoke, the hatred and the military involvement in civilian affairs ceased and peace once again looked as if it would succeed in a bolder and more permanent manner.

Dudley had pulled some "fast ones" during the infamous Dudley Board of Inquiry held at Fort Stanton and had tried to justify some of his actions. His prejudicial treatment of one side over the other was obvious. He brought cannon into Lincoln, and fired into the McSween side's homes without remorse, or so it seemed. It was obvious to many, including the widow McSween, that Dudley was under the influence of the Murphy/Dolan gang, and she fought hard against the man, to no avail.

Nevertheless, when all was said and done, Wallace, by transferring most of those officers who were in charge of certain disputes between soldier and civilian out to other posts, created among the locals a much easier feeling about resuming good relations with the Fort.

Literally hundreds of books and articles have been written about the Lincoln County War, and Billy the Kid. More will be written but the wild frontier known to Billy and Pat Garrett and Victorio was quickly coming to an end. In fact, the Lincoln County War and the Apache Wars both ended within a few years of one another. As the Apache Wars wound down, uses for military forts throughout the southwest were downgraded and most were abandoned. When

Courtesy Lynda Sánchez Collection.

One of our greatest and most memorable generals received part of his military training at Fort Stanton. As a young man Pershing was assigned to many of the Forts on the frontier including Fort Bayard and Fort Stanton. Although these assignments were toward the end of the frontier era, the land was rugged and times primitive by any standard, and called for men of great fortitude. His time in the West honed his skills and forged the young lieutenant into the brilliant and tough general in command of the American Expeditionary Forces sent to fight in World War I. He had also spent over a year chasing Pancho Villa after his audacious attack on Columbus, New Mexico in March of 1916. This became known as the "Punitive Expedition". They scoured the Sierra Madre chasing Villa and his rag tag band of men; they developed maneuvers for future military forays and used planes for the first time. It was good training for the men, but very frustrating for the good general as they never caught Villa.

Nevertheless, despite a lack of supplies and poorly trained and armed troops he brought order out of chaos (in Columbus)… *"Pershing would not brook disorder or sloppiness of mind or person or billet, and he was a superb organizer of troops…every unit…every horse and every man was fit…discipline was perfect (at least according to Patton)… "He was an imposing figure, tanned, ramrod-straight, and meticulously groomed. This man had presence, and he was all soldier."* (D'Este p 165, 189)

Another imposing and fearless general during World War II was also trained and mentored by Pershing. That individual was General George Patton, his aide-de- camp for several years during the Punitive Expedition into Mexico and World War I.

No matter the assignment or where he was in the world, Pershing succeeded. During World War I he was *"charged with one of the most challenging tasks ever given to an American military commander: to create and train a fighting force from a tiny army that was ill-equipped to fight any kind of war…"* (D'Este, p 189) But just like in New Mexico, when chaos reigned supreme before his arrival in Columbus, he knew that everything would be carried out according to his wishes and victory would ultimately be at hand. Yes, the frontier West helped make the man and many who are proud of the history of Fort Stanton, point with particular pride in their knowledge that Black Jack Pershing along with the Buffalo Soldiers were stationed on site.

Black Jack Pershing

Courtesy Lynda Sánchez Collection.

the time came to muster out, innumerable soldiers decided to remain because they observed that Lincoln County was progressing and leaving the outlaw stage of its existence. . They married local Hispanic women and their descendants remain as part of the fabric of the territory in terms of traditional and modern life style.

One old timer quipped recently, *"Yup, 'twas bugles, bullets and brew once. Now, 'tis pretty quiet. Them buildings would tell quite a story, and I hope you all tell it true!"*

Dedicated Fort Stanton volunteers are working to collect more of the fort's legacy and to house it in a new museum that will soon be refurbished for the artifacts, documents and history of the area so that they can *"tell it true!"*

It is peaceful now, with only the thunder that comes from our New Mexico summer rains bringing "*la lluvia bendita del cielo*" (blessed rain from the heavens) for all who live along the base of the sacred mountain, Sierra Blanca. The thunder clouds of war and conflict are no longer shadows of our past.

Photo by David L. Tremblay

Sources and/or Recommended Reading:
(Also see sources at end of Chapter I)

Bilbo, Mike, numerous publications…contact BLM office, Roswell, New Mexico

Billington, Monroe Lee, New Mexico's Buffalo Soldiers, 1866-1900, Niwot, Co,: University Press of Colorado, 1991

Recorded in the Capitan Mountaineer, December 15, 1923, Ms. Karen Mills, Lincoln County Records Clerk, sent the interview to me as a point of interest. Corrections are noted in ( ). 2008

D'Este, Carlo, Patton, A Genius For War, New York, Harper Collins, 1995 (pages 154-164; 189; 315-320)

Fulton, Maurice G., History of the Lincoln County War, edited by Robert N. Mullin, Tucson, University of Arizona Press, 1968

Leckie, William H., The Buffalo Soldiers: A Narrative of the Negro Cavalry in the West, Norman, University of Oklahoma Press, 1967

Myers, Lee, Fort Stanton, New Mexico, The Military Years, 1855-1896, Lincoln County Historical Society, #2, July, 1988

McBride, James J., Interned, (Internment of the SS Columbus Crew at Fort Stanton, New Mexico, 1941-1945), Santa Fe, 2003 (self-published)

McBride, James J., A Photo Essay History Fort Stanton Marine Hospital, Public Health Service 1899-1953, Santa Fe, 2005 (self-published)

Nolan, Frederick W., The Life and Death of John Henry Tunstall, Albuquerque, University of New Mexico Press, 1965

Ryan, John P. Fort Stanton and its Community 1855-1896 Las Cruces, Yucca Tree Press, 1998

Simmons, Marc, Stalking Billy the Kid, Brief Sketches of a Short Life, Santa Fe, Sunstone Press, 2006

Thompson, Jerry D., Desert Tiger, Captain Paddy Graydon and the Civil War in the Far Southwest, 1992, Texas Western Press, El Paso

"On the web"

www.civilwarcavalry.com | www.Musketoon.com

www.freerepublic.com/focus/f-vets | www.news-bulletin.com/lavida

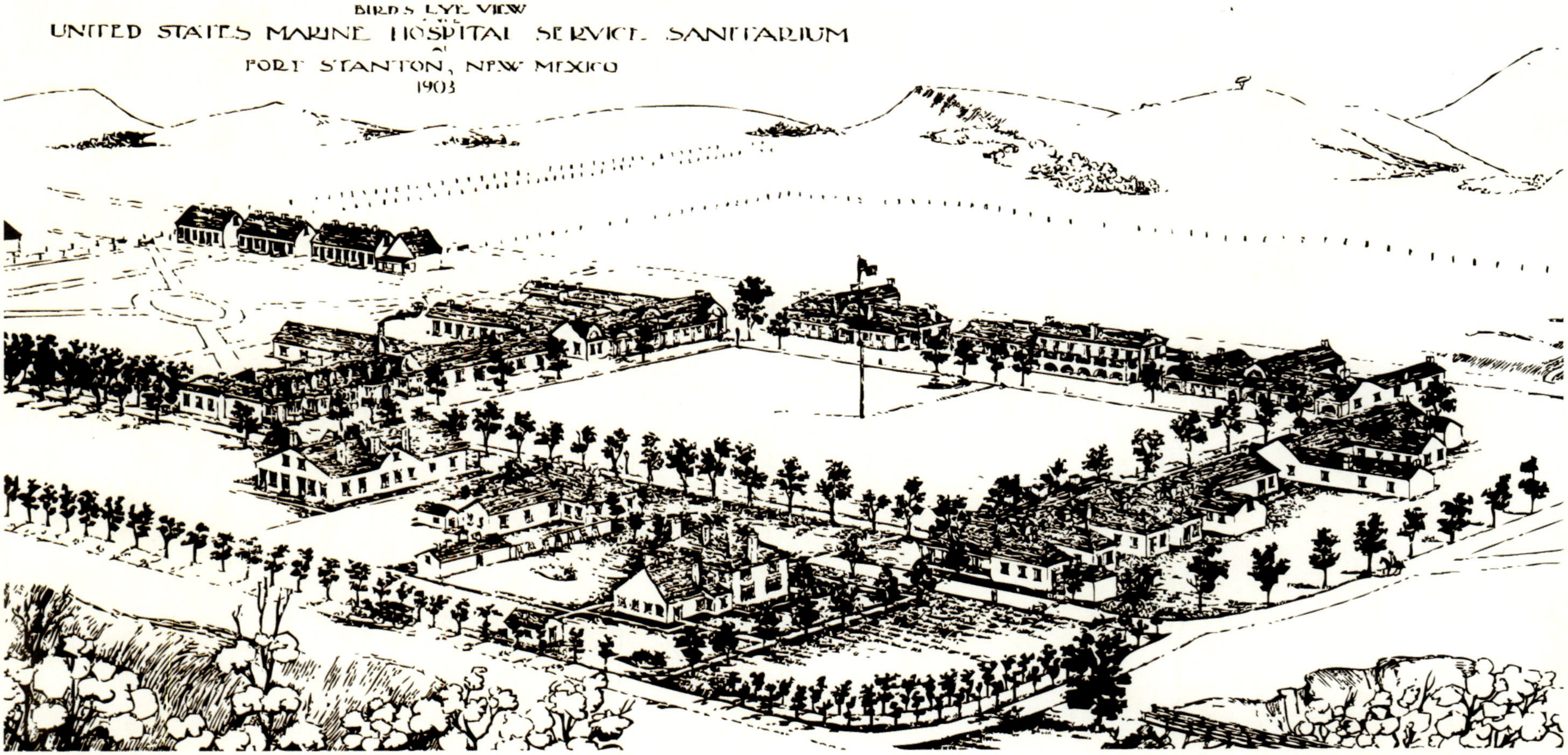
BIRD'S EYE VIEW
UNITED STATES MARINE HOSPITAL SERVICE SANITARIUM
AT
FORT STANTON, NEW MEXICO
1903

1896 painting of Sierra Blanca and the Fort in Winter, by Irwin Murdock. Courtesy Bob McCubbin and Michael Vinson.

Chapter 3

The Public Health Service, 1899-1953

(Conquering the White Plague)

The military purpose for which the fort had been constructed changed in 1896. While the post was no longer needed in a military sense, its legacy was to move on from peace keeper to yet another tradition, that of healing, when, in 1899, the Fort transitioned into a tuberculosis hospital for Merchant Marines. This new tradition of healing also continued the economic development of the district for more than another century.

New Mexicans understand the outlaw and Apache conflicts and the ranching and mining industries. However, the majority are totally unaware of the influx of people created by the health industry and the health seekers who arrived at the turn of the century. (Laws; v-vii)

Tuberculosis or the "White Plague" affected the settlement of the West and should have its rightful place in history as a major reason for immigration westward. Most people do not realize that during the 19th century, it was the leading cause of death among people 15-45 years of age. In addition to gold and land seekers what is not publicized is the dramatic and often desperate immigration into the arid southwest of health seekers. By the mid 1850's *"altitude therapy"* was heralded as a new form of treating tuberculosis. A flood gate was opened and literally thousands came into New Mexico and to Fort Stanton once the fort was converted to a sanitarium for Merchant Mariners that had contracted the disease.

On April 1, 1899, by way of a presidential executive order, President William McKinley established the Fort as the nation's first tuberculosis hospital. The previous year Dr. J. O. Cobb had searched the Southwest for a location, and he found Fort Stanton to be ideal because of *"its ready water supply, mild climate and restful environment."* (McBride, Fort Stanton Marine Hospital, p 6) The Fort soon became a model of healing and care for hundreds of patients over the next several decades.

The first patient to arrive on November 18, 1899 was Seaman Joseph Ryan. Five months later more than 65 Merchant Marines had been admitted. Twenty eight years later the fort had over 236 patients. A

Fort Stanton aerial shot, 2009

Photo by David L. Tremblay

library of 3000 books was well used and patients could do whatever they liked providing it did not involve too much exertion. The Knights of Columbus showed movies two nights a week and dances, plays and other entertainment were much appreciated. Churches, baseball, a Bridge club, an orchestra and even a Fine Arts Club were organized and enjoyed for many years.

Memories, too, of Merchant Mariners sunning themselves and improving their health are ever present in the minds of many area residents. Their parents and grand parents worked at the fort. Once cured of the dread disease, many had come to love the area and settled nearby valleys and communities, as had the soldiers before them.

One of these individuals suffering from the dread disease happened to be a physician himself who ended up at Fort Stanton. Dr. James W. Laws arrived in 1902 from Tennessee. Cured *"of the disease by a stay at Fort Stanton, he remained in the area, marrying another health seeker, Grace Austin...."* (Laws; vi)

His story is similar to so many who came to be healed and stayed for a life time.

Because of his medical background, he had inquired of the post commander if there was any employment available at the fort. Eventually he was told to report for duty. His less than auspicious

U.S. Army Archives

Fort Stanton aerial shot, c early 1940's

Lynda Sánchez Collection

Tubercular Patients – c 1920's

arrival at the fort is a vivid description in his memoirs:

"I had to wait until the afternoon before the mountain wagon came over from Fort Stanton. The wind was blowing so hard that I thought it wise to fish out an old hunting hat....As we rounded a curve and entered the Bonito Valley, the old cavalry post, Fort Stanton, came into view with the snow capped White mountains framing the upper portion of the picture and the fort itself, surrounded by rounded foothills covered with gramma grass and here and there bunches of Hereford cattle grazing on the brown winter forage. ....I learned to accept this picture, with the snow capped White Mountains in the west and the blue El Capitan mountains in the northeast as a rather attractive setting, but in my present mood, after a night of broken sleep, I kept saying to myself, "I will try to put up with this Godforsaken country until I recover my health, and then I'll shake the dust of it from my feet and depart without regret." (Laws; 7)

Needless to say, Dr. Laws changed his mind and became a permanent resident of the area after he was cured of the "white plague". Establishing his own sanatorium in Lincoln about 8 miles down the road from the Fort, he had regained his health and mission in life. And this was so often the situation with individuals from all walks of life who arrived in Lincoln County. They came to love the area and wished to contribute to its growth and future.

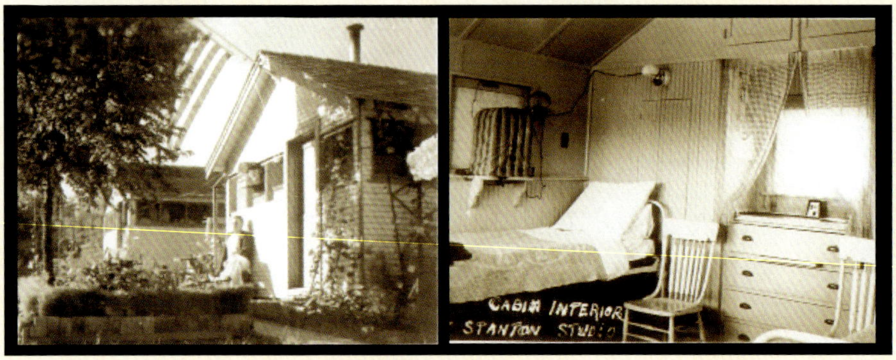

Left: A typical patient cabin
Right: Cabin interior

Courtesy Fort Stanton 'Kids' Photo Archives.

Patients at the Fort were primarily sailors with the Merchant Marine Military Service. They were a tough lot and often confounded the locals with their own brand of the English language. For example, when they left the Fort for a visit to Capitan, they spoke of the trip as "going ashore". Or they would describe other events in seaman's language. "A lady passed here a few minutes ago with a sou'wester made fast to the stern of her saddle". It took a while to get used to, but eventually communication was not so difficult. Rows and rows of tents were set up for the influx of patients. Diversion came from dances or trips to local communities. Nevertheless, it must have been difficult for those salty sailors to live in the wilds of *waterless* New Mexico!

The seamen were also expected to pay their own way to New Mexico. They were basically given directions to the Fort, and the required papers for admission. Upon arrival at the isolated outpost the patient would go through medical exams, receive a set of regulations and thus was the poor man's introduction to the life of a tubercular patient.

The rules were strict and each man was expected to follow them to the letter. They were expected to rest, eat well, exercise when possible, keep their abode clean, keep a cheerful disposition, and spend as much time as possible breathing the arid and fresh air of New Mexico. For a good long while during those first two or three decades they could not drink, play cards if they had a fever, and had to bathe at least once a week. The men could only spit in their *"sputum cup"*, use water with great care, not waste anything, could not drink out of a milk bottle, respect the flag at all times, and they could not own an automobile unless special permission was granted.

Another point that must be remembered is that the first two to three decades were very difficult in terms of amenities and discipline for <u>both</u> patients and staff. However, from the earliest years the staff enjoyed steam heat in the Parade Grounds buildings and indoor plumbing. A luxury no doubt for most people had neither in other areas of the County. Life was also harder in many ways compared to the decades after prohibition was lifted. Additionally, modern construction and supplies began to filter into the Fort just prior to and then after World War II.

The Little Bonito, Child of the Snow"

(Published in the Carrizozo Outlook, August 22, 1913, the information and poetry was written by a visitor to the Fort, P. G. Zimmerman. Ironically, for all his discussion of peace and tranquility, one year later, 1914, World War I broke out in Europe. Yet his descriptions fit Lincoln County today just as much as in that era almost 100 years ago.

The sacred mountain, Sierra Blanca, looms tall and proud dominating the landscape for miles and miles. Sacred to the Apache people and to all of us who care about water, land and an interconnection with ourselves and mother earth, this poem is most appropriate today as we struggle with our desire for growth and our hope for the future that must encompass respect for our homeland and for water.

> *"This poem was written for my little friend and comrade, Mona Lane, and is dedicated to the living and dead of Fort Stanton, New Mexico.*
>
> *"The Little Bonito is one of those pretty little mountain streams that rise amid the banks of snow in the White Mountains. Hence we call it 'Child of the Snow'.*
>
> *"Fort Stanton is a government Marine hospital and Sanitarium, and nestles on the banks of the Little Bonito 22 miles from the White Mountain. It was a military post from 1857-1897. The post now has a large herd of cattle, some horses, and mules, dairy farm and orchards and things too numerous to mention. The entire valley is irrigated from the Little Bonito which is verily the life of the valley.*
>
> *"I spent the 20 and 21 of May in this post. Met many of the officers and people of the Fort. Dr. Smith is commander-in-chief, and at present there are some 200 patients in attendance. Seeing those sick fellows in the reclining chairs, sitting in the warm sun in the afternoons, as they watched in dreamy retrospect the silvery tides of the Little Bonito as they trickled by was the inspiration of this little poem."*

The Little Bonito, Child of the Snow

Flow on thou Little Bonito,
Scatter gladness wherever you go;
Fresh from thy home in the mountains;
And the beautiful drifts of snow.

Bear your silvery floods to the flowers
That so wildly bloom on the plain,
Be you their sweet soul and comfort,
Be you their sunshine and rain.

Come trickling down through Fort Stanton
'Mong the sleepy villas and trees;
Bless the tired hearts of the patients
That recline in their chairs of ease.

For they shall list in the evening
To your music so soft and low,
And in their hearts they will bless you
Chiquito Bonito, sweet waif of the snow.

Your course is so mild Bonito,
In your flight to the southern sea,
That pilgrims touched with your beauty
Have come to Live and die by thee.

They've come from the storm swept ocean;
The dark and heaving starless mane.
They've come to rest from the struggle,
Secure from life's worry, and pain.

Sing to them sweet Bonito;
Sing of the days so long gone past,
'Fore your waves were jarred by the drum,
Or stirring bugles war-like blast,

Sing again of a day to be,
When wars shall drench no more the sod,
When earth shall be one brotherhood;
The peaceful home of man and God.

(One year later, World War I began)

Author's note: The document was forwarded to me courtesy
Ms. Karen Mills, Records Clerk for the County of Lincoln.

oreground photo courtesy Fort Stanton 'Kids' Photo Archives. Background photo by David L. Tremblay

Nurses Quarters now, and then.

Photo by David L. Tremblay

Courtesy Fort Stanton 'Kids' Photo Archives.

Courtesy Fort Stanton 'Kids' Photo Archives.

Fort Stanton's first fire truck, 1916.

There were massive construction projects beginning in 1934-35 with construction of the *"new"* (present) power plant, demolition of the old power plant that dated from the military days, construction of the "new" hospital on that site (it opened in May 1936), then demolition of the *"old"* hospital. The construction of this new and updated hospital opening in 1936 was a big morale booster for all concerned. A unique feature was the first "electric" elevator in New Mexico! The majority of this construction was essentially finished by 1940. Then came World War II and rationing created more hardships for some, but food never really was a problem because of the ranch and farm on site.

Fire danger was always a continual reminder of the age and condition of the structures. According to Danna Kusianovich Henderson, *"The buildings were numbered for fire protection purposes, starting with #1 (C.O.'s house) up to #14 around the Parade Grounds. The fire whistle blew one long for 10 and short blasts for 1–9, so a long and three shorts would send the guys with the fire cart (which was pulled by hand) to #13. We grew up with that system, so to us the buildings are always identified by their numbers Actually; I believe that numbering system started in the military days for the same reason."*

A cemetery was also established nearby taking the place of the old Cavalry cemetery on the hill. Today approximately 1700 merchant mariners, armed forces veterans, their families and staff from the fort have found this cemetery as their final resting place. Hopes for the cemetery to be expanded and to continue including veterans from all branches of the military are now being considered. The peace, beauty and calm of this setting serves as a special tribute for our nation's veterans. Under the leadership of John Garcia, Cabinet Secretary for Veteran's Affairs, such a transition could become a fitting example to perpetuate the legacy of honor for which the Fort is known.

Photo by David L. Tremblay

Fort Stanton Merchant Marine Military Cemetery.

Courtesy Danna Kusianovich Henderson.

Fort Stanton 'Kids' all grown up.
Front (l-r) Mary Fish Manies, Claire Boyd Fluharty, Tom "Skip" Gould and Danna Kusianovich Henderson
Rear (l-r) Luke Manies, Arv Fluharty, Jimmie Thompson, John Kusianovich

GROWING UP PROUD...

Because there were so many families on site during the latter Public Health Service (PHS) years, (1920's to 1950's) two very separate worlds existed. Although they often ran parallel to the care givers and the hospital's purpose, that of healing wounded warriors and helping them to contend with one of the most dread diseases of our time, the world of these youngsters *"growing up proud"* came to represent another aspect of life at Fort Stanton.

Among the dozens of children who grew up at the fort and the families that nurtured them, a common theme among the vast majority was of freedom, respect for the land, adventure and rarely being bored. Some might have been a bit lonely, however, overall, there were chores, but also plenty of time to run, jump, hike, explore, fish, hunt, watch movies, lope horses across vast areas, and attend the famous Fort Stanton rodeos. The children learned from the patients and staff many life skills or about other parts of the world and the cultures therein.

Tom Gould noted that, *"Merchant seamen came from many different parts of the world and we kids were exposed to a diverse mixture of cultures. At our house we had a Ruth Roll from Norway stay with us for a long period and I thought of her like an Aunt. Her husband was a captain who stayed with his downed ship…He was permanently disabled by the event and hospitalized at Fort Stanton. Aunt Ruth spent hours telling me about the Vikings, and that during thunder storms Thor was hurling his lighting bolts…*

"Spiro Vastniny always made the best spaghetti and insisted the Greeks, not the Italians, invented it. Of course another of our frequent visitors, Italian Mike Nero, would always strongly debate that. The point is, that as we kids grew up we were exposed to many cultures, an experience that was unique and probably impossible to repeat. That experience has served me well when I went to school in Mexico and traveled in South America. It also helped me win the hand of my beautiful Peruvian wife so many years ago."

(Gould, Tom, email, April 15, 2009)

Courtesy Danna Kusianovich Henderson.

Dan Kusianovich with Peter Hurd again reiterating the art and culture that was available and influential in their lives.

Danna Henderson recalls learning about the geography of the world during World War II from a map her father kept on the wall pin pointing sites of battles and where friends or family were serving, or where and how relatives and friends in the Old Country were being affected. Therefore, education was an ongoing twenty four hour a day project, not just sitting in a class room.

Finally, when they reached adulthood and were ready to head out into the world, the majority departed as strong, gentle, successful and proud individuals. During interviews with the author, they spoke fervently about their growing up years. These were not memories painted by rose-colored glasses, for they all admit times were tough, especially during the war. Also, during the early years of the hospital, the lines of demarcation between staff on the quadrangle and elsewhere were specifically observed. The children felt the segregation, but less than their parents who had to adhere to the regulations. Even that line became blurred over decades.

Nevertheless, the events, the healthy environment and the love of devoted parents and examples of healing and care giving, of livestock husbandry, interesting clients and World War II victory gardens, never left them. Nearly every child who grew up at the fort, went on to graduate from college, forging impressive careers for themselves.

QUOTES FROM SOME OF THE "CHILDREN"

"I think of the Navy pilot who got lost… from San Diego. He just knew he was in Mexico and what was he going to do? He saw an American flag and brought his plane down on the field… an American flag and that was good enough, but they had to dismantle the plane, haul it out with trucks, and I'm sure he lost his rank."
Ervin Aldaz, from speech transcript, Lincoln, 1997

"The World was great and wonderful and I was busy enjoying the mountains and the freedom… **Jimmie Thompson**

"My memories of Fort Stanton are warm, happy ones…We had a fantastic Boy Scout Troop with Ray Baker, the Scoutmaster. He made sure we did a lot of camping, hunting and studied outdoor lore. The lessons I learned then have helped me more times than I can remember… **Tom Gould**

"The Blue Gorge up the Bonito River was a favorite place…with steep rocks and pools of water where we could play. The whole community raised the children at Fort Stanton. We were trusted, loved and protected. Not bragging, just blessed." **Eva (McCallum) Howard**

"A Hard Landing" Beechcraft Model 18 C-45 c.1940's

Courtesy Danna Kusianovich Henderson.

Courtesy Danna Kusianovich Henderson.

Arrival of "Jenny" Curious patients and staff, c 1924.

"... summers were best. Freedom and adventure came with summer...I treasure my memories of growing up in Fort Stanton. It seems that it was so wholesome and simple...Even now...when I let my mind drift back to those days, I ....relive some happening or other, walk the hills, visit with friends, lie on my back and watch the millions of stars in the sky. Then I come back to my present world refreshed and grateful."

Margaret (Searle) Howard

"Growing up in Fort Stanton was an experience all its own. In retrospect, we were a privileged group......the children of Fort Stanton were pampered yet disciplined...they did not need someone to plan their activities. They created their own agenda...One rarely heard the words, "Mom, there is nothing to do." We had a playground that covered almost 49 square miles. We roamed the hills and canyons in search of new adventures and we had our favorite sites to visit on a regular basis. Most of us also enjoyed the library and were avid readers...

"A youngster with imagination and initiative could find many exciting things to do and places to go. It was the same with our parents..."

Eugene A. Merrell

"You mentioned our 'isolated' part of the world, but that wasn't really the case at Fort Stanton, which made it a unique place to grow up. The merchant seamen were from all over the world, literally, so we were exposed to many different cultures. The doctors were there for two or three

The First Commandant's House, later Nurses Quarters.

Courtesy Fort Stanton 'Kids' Archives.

years, often came from big cities, and brought that brand of culture along. We had constant visitors from Washington. My father, being an artist, was great friends with the Hurds and through them the Wyeths, Paul Horgan, etc. He also had good friends at the El Paso Herald Post and knew Ernie Pyle and Bill Mauldin.

"We also often had officers from Walker AFB as visitors, so we may have been isolated geographically but not culturally. We sure knew a lot more about the world than today's teenagers!! Our families also traveled a lot before and after the war--my folks took me to the 1939 World's Fair in San Francisco, we visited back and forth with relatives in Galveston and Miami... no, the Fort was a special place."

Danna (Kusianovich) Henderson (May 13, 2008 email)

... "it was nice to live so close to school and to go home at noon for lunch. The teacher let my dog come in and sleep under my desk..."

Billy Dale Brown, 1940's

"My Shangri La does not exist in the faraway Himalayas, as you might imagine, but in the Land of Enchantment located between the Sierra Blanca Mountains and the Capitan Mountains at an elevation near seven thousand feet. Old Fort Stanton was Paradise for me."

Dr. Albert Rood **See Chapter IV.**

Another example woven into their memory was winter's magic

Courtesy Fort Stanton 'Kids' Archives.

An icy path along the Parade Grounds, c 1930's.

and the coming yuletide season. There was just something about winter's white and graceful beauty. The patients, staff and the families with children watched as the blazing colors on the Sacred Mountain, Sierra Blanca, were replaced by the icy fingers of winter and the time to rest for creatures, great and small. All the earth was at rest except man, of course.

Nostalgia abounded during that special time of year as families prepared for the coming holidays. However, it was a simpler time when the children who grew up at Fort Stanton and other parts of Lincoln County also went out with parents to bring in that special tree, or hunt for turkey and venison or make snow men on the wonderful parade ground right in the middle of their lives and homes.

Christmas trees were fresh cut and the flickering lights dancing among the branches were really tiny white candles. Great care was taken so as not to cause a home fire. Everyone was vigilant and a bucket of water was kept nearby. Later, electricity came in and the people converted quickly to the easier and multicolored lights. However, many recall the special glow of the candles against the green of a simply decorated tree as being unique.

Winter in Fort Stanton, c 1930's.

Courtesy Fort Stanton 'Kids' Archives.

Most families cut their trees on Christmas Eve day, and certainly would have been horrified at the over commercialization of our modern world with Christmas "everything" being put out well before Halloween. Old timers just shook their heads in sadness but they smile when they describe their own distinctive memories. Gifts were basic and of a practical nature. Most were hand done and definitely not of the plastic mode that so many of our children are inundated with today.

During those years families were more prevalent than during the military years, and children were everywhere. A school was established on site and in general this unusual community was self-sufficient and provided work for most of the county. The classic Victorian "officer's quarters" had been converted into homes for the Public Health staff. Best of all, the beautiful bay windows were the perfect place for Christmas trees.

At Christmas there were numerous fiestas and the ones held in the Community House (where the Fort Stanton museum is located today) were the most fun for the children. Santa arrived bearing gifts of candy and goodies for every child no matter the age. In private homes the tree was awaiting Santa's arrival for not only did old St. Nick bring presents, he *decorated* the tree in some homes, even as he filled all stockings before he departed! Children received gifts such as a doll, marbles or rifle but mostly they received practical items like socks and flannel shirts, a nice dress or books. The children were, much as today, awaiting the arrival of the jolly gentleman from the North Pole with great anticipation! That never seems to change.

Jugs of homemade elderberry wine were enjoyed by the adults while children played outside in the cold and snow or went riding in wagons or sleighs. Though a few had cars it was not so easy to get around slick roads in automobiles. And so it was every Christmas around the elegant parade ground and in the homes or churches at the Fort.

Author's Note: One could go on and on with tales about the Fort. It is suggested that the McBride books listed at the end of this chapter be consulted for more stories and adventures of the *"Children of Fort Stanton."* And, as with Jim McBride, I have found that the two proud and determined ladies who began taking the information from the old timers, who encouraged others to write their memories and to send in their photo documentation, have also become my friends. Danna Henderson and Claire Fluharty are amazing women in their careers and especially in their dedication to seeing that the post cavalry era is documented and not forgotten!

Vintage Collection, IMSI

The Spirit of Christmas, 1931

Willett White, one of the children who grew up at the Fort described this touching Christmas story that made the children appreciate their own lives and homes. It happened in 1931, two years into the Great Depression:

"It was Christmas and everyone was awaiting the arrival of Old Saint Nick. Children were excited because they knew that gifts would be delivered and a new year would be coming soon. However, this year a delay caused Santa to be later than usual and everyone was wondering why...

"Someone had alerted the Fort staff that a vehicle and a group of several adults and children were camped just below the Fort along the Bonito River. They were cold, hungry and shabbily dressed. Their truck would not start but they promised to move along as soon as they could get it repaired. Apologizing, they explained that they did not know they were on post property. They had been there for a couple of days and had not expected to be camping out in winter.

Once discovered by the Fort Stanton folks, their tragic plight changed rapidly. Blankets were sent down to the people and hot food was delivered. Supplies were provided for them for the rest of their long journey to California and Mr. Burleson, brought the truck to the Fort garage for repairs. But, what could they do about those kids with this being Christmas? Somehow Santa had to help out too.

"The children were told that Santa had not expected to find so many children at the Fort, and he was hoping they would understand the need to share some of the toys he carried. Thus, Santa was able to visit the encamped group delivering Christmas toys and good will to some very happy children and parents that no doubt had tears in their eyes, but they were tears of joy that Christmas day.

"Soon they moved on to California as the precursors of the thousands who traveled from the Midwest during those dreadful dust bowl years. However, that group of travelers continued on with the hopes and best wishes from Fort Stanton residents for a safe trip."

Courtesy Danna Kusianovich Henderson.

Dan Kusianovich at his desk

THE HOSPITAL TRADITION CONTINUES, 1953-1995

THROUGH ALL OF THESE DECADES THE FORT HAD BECOME A VERY INDEPENDENT RURAL COMMUNITY. THERE WERE CATTLE, CROPS AND EVEN DAIRY BARNS ON SITE.

Courtesy Danna Kusianovich Henderson.

Ms. Claire Boyd Fluharty enjoying a summer day with friends on the Rio Bonito.

Ms. Claire Boyd Fluharty, daughter of the ranch/farm foreman during the 1930's to 1950's has described her life as *"one of challenge, fun and character building. It was not lonely as some may have thought, and an individual developed a lasting independent spirit."*

The Fort remained a stalwart center for commerce, health care, employment and social activities. Children and the elderly, some from as far away as Mescalero, were often brought for visits to the dentist or doctor at the fort .The small villages of Lincoln, San Patricio, Hondo, Glencoe and others had none. Many old timers fondly remember dances, baseball games and just *"visiting"*. Rodeos were popular and many cowboys came from afar to compete at the rodeo grounds. Today only silence greets one at the corrals and chutes where once lively competitions occurred. Danna Henderson recalls competing and running barrels. And, of course, she loved to ride the hills surrounding the fort.

Danna Kusianovich Henderson, on horseback at the Fort Stanton stables, 1947

Courtesy Danna Kusianovich Henderson.

Peace after World War II had brought relative calm to the nation, and the dread disease of Tuberculosis or the "white plague" (white death) was finally conquered. No longer were the Merchant

Dan Kusianovich and Navajo patients.

Marines sent to the fort because the clear, arid conditions of New Mexico had done their job. However, in New Mexico the Native Americans still had major problems on their reservations so the fort was transformed into a state hospital for treating these cases of tuberculosis.

Apparently the US Indian Health service had contracted with the Fort. One of the major figures of that era was Annie Dodge Wauneka, daughter of the famed Navajo leader, Chee Dodge. She was the first woman elected to the Navajo Tribal Council and was awarded the Presidential Medal of Freedom by President Kennedy for her public health activisim and battle against tuberculosis among the Navajo. Thus for several years the fort's clients were primarily Navajo.

There were some great stories about the multicultural aspects on site. Ms. Henderson recalls *"a medicine man making the rounds with the doctors each day. And when the first patient died, all the Indian patients moved out of the hospital and camped on the Parade Ground until someone (maybe Annie) convinced them it had been cleansed of evil spirits and talked them back inside…"* (Danna Henderson, email, April 14, 2009)

The fear of Chinde or witches and ghosts is still, as of 2009, prevalent among Navajo and Apache people, especially tribal elders. Clients were also flown in from the reservation, so probably were transferred from one of the hospitals in Gallup or Fort Defiance.

Perhaps Danna Henderson, who summed things up as events coming full circle, is correct when she wrote recently:

> *"The eras (of history at the fort) actually are intertwined, and go full circle if you think about it. You start with the Native Americans and then the military arrives to "tame" them (along with trying to tame the lawless frontier epitomized by the Lincoln County War), and then another branch of the military, the Marine Hospital*

Courtesy Danna Kusianovich Henderson.

Navajos gathered for a film in a hall at Fort Stanton.

Service, arrives to tame the "white plague." Merchant seamen arrive from all over the world, creating a uniquely cosmopolitan town in what wasn't even a state then. They marry nurses who came from all over the U.S., and remember that most of the "children," are children of those patients. Doctors come for two to four years to fulfill what amounts to military obligations, bringing a "big city" atmosphere with them. Meanwhile, many of the patients are getting to know and becoming friendly with the Indians. My father not only made marvelous drawings of many of them but worked closely with people like Annie Wauneka. And when the PHS abruptly pulled out in 1953, Fort Stanton became an Indian hospital for a period of several years completing the circle."

(email, April 14, 2009)

Innumerable individuals interviewed have commented upon the abrupt closing of the fort by the federal government. As with so many situations of this kind, the rumor mill was running rampant. First the post was to be closed, no funding was appropriated; then it was to remain a while longer even though tuberculosis was now being cured by antibiotics and other modern medicines. And then it happened, abruptly, secretly and with an air of no appreciation for the great work done by the Public Health Service doctors, nurses and staff for decades!

> One individual wrote recently, *"When the US government quickly decided to close Ft Stanton, it was the most traumatic event in my life--and in the lives of many others. (Now, years later,) I sincerely believe it needed to be done, as TB COULD be treated well with medication. But the affect on the people working there--people who had been promised during the war that if they stayed and didn't leave for a better job at the time--the government would stand by them afterwards."*

Unfortunately, as with the 1990's closing of the fort as a hospital for the mentally handicapped, many people were scattered to the four winds, jobs were down sized or even lost and a mini-depression hit Capitan and the surrounding areas of Lincoln County.

It was the opinion of many that the state ran a poorly managed system with all workers who stayed being paid less and treated quite differently. So, most people were hurt and angry. Many never saw their closest friends again. The financial difficulties were immense. The salaries at Fort Stanton had been the envy of the local people--particularly because of the benefits of cheap housing and food. But in the outside world, the salaries were really low. So to claim it transitioned to the state as though it was just another logical, easy change was incredibly hard to accept.

> Claire Fluharty wrote... *"For me, I felt like I changed planets... My father (farm foreman) struggled a lot, trying to make the fort ranch pay. He was not allowed to sell anything produced there*

locally, as it would be competition with private enterprise. The days of Carville, Louisiana (where the fort had supplied meat and dairy products) were gone, and so everything was cut back..."

After about five years the Indian hospital closed. The fort was then transformed into an institution for the mentally handicapped and until its closure by Governor Johnson in the mid 1990's for "economic" reasons the employees tried to ease the care giving for parents of mentally handicapped individuals. The medical care and educational programs were attuned to these needs. A greenhouse was established as a source of income for the clients and once again, many of the locals were hired to work at the fort.

However, along with this period, unfortunately came abuse of the structures. The Victorian or Pueblo Revival styles mattered little because money never seemed to be available for even basic maintenance. Slowly Fort Stanton began to slip into a sad state of neglect, *(as noted in Fred Nolan's introduction)*. Yet the staff soldiered on despite this lack of interest from Santa Fe. Changing philosophy in caring for the mentally handicapped, and the economics of a large staff and few clients were just too expensive to continue. Many clients were "main streamed" back into the world from which they came. In the mid 90's the post was finally closed as a hospital. It was converted to a women's prison, and later a rehabilitation center for drug and alcohol abuse.

The tradition of healing and hope continues today but with another goal in mind - that of preservation of the entire site including the historic quadrangle, the stables, the German Internment camp, and the cemetery in conjunction with the Bureau of Land Management property. They administer over 25,000 acres surrounding the fort, adding to the mystique and atmosphere of an area little changed in the last century. It provides a wonderful buffer zone that aids in the preservation of the heritage of over 150 years of public service to the people of the United States and New Mexico. The stalwart structures now have an opportunity to once again be refurbished into buildings of grace and elegance from an era long since passed. We shall see what the future brings.

Photo by David L. Tremblay

Hand-made tile sign made by the resident children of the School for the Mentally Disabled at Fort Stanton (1960 - 1995).

Sources and Recommended Reading:

Kashima, Tetsuden, Dr., Judgment without Trial (Japanese Imprisonment during World War II, Seattle, London, University of Washington Press, 2003

Laws, James W., M.D., Memoirs of a Country Doctor; (edited by Ann Buffington), Series 1, Lincoln County Historical Society)

Myers, Lee, Fort Stanton, New Mexico, The Military Years, 1855-1896, Lincoln County Historical Society, #2, July, 1988

McBride, James J., Interned, (Internment of the SS Columbus Crew at Fort Stanton, New Mexico, 1941-1945, Santa Fe, 2003 (self-published)

McBride, James J., A Photo Essay History Fort Stanton Marine Hospital, Public Health Service 1899-1953, Santa Fe, 2005 (self-published)

McBride, James J., Editor, Children of Fort Stanton, self-published, Santa Fe, 2008

Rood, Albert C., MD, Medical Mysteries in a Land of Many Cultures, private essay, 1973, courtesy of Ms. Danna Henderson.

Ryan, John P. Fort Stanton and its Community 1855-1896 Las Cruces, Yucca Tree Press, 1998

Interviews, Correspondence and emails:

John J. Corcoran (Fort Stanton Cave Research Team) interviews, emails and reports 2007-2009

Claire Boyd Fluharty, many interviews and emails, 2008-2009

Garcia, Phil, interviews, 1978-1980

Tom (Skipper) Gould, many interviews and emails, 2008-2009

Danna Kuscianovich Henderson, many interviews, emails and correspondence, 2008-2009

Kashima, Tetsuden Dr., telephone interview, June, 2005, regarding Fort Stanton future

Vintage medical. Composite by Laura Reynolds. Images: phisick.com and dreamstime.com

Photo by David L. Tremblay

Wild antelope grazing near Fort Stanton.

Chapter 4

A Doctor's Life at Fort Stanton

After obtaining a fishing permit from the Indian office of the Mescalero Apache reservation we then rented a rather tricky old mule at Alto and packed our tent, grub and other camp gear on the beast. We began a six mile hike into the mountains near where the Ruidoso ski run is now located. Slim led our pack mule and I walked behind as a precaution, but even so halfway up the trail our pack animal jerked the halter rope away from Slim, quickly circled around me through the timber, and hightailed it for home...

There were many poor young families with babies in Lincoln County, especially in the towns of Capitan, Lincoln and San Patricio, who rarely had medical attention, so I organized a well baby clinic one day a week to try to improve the situation. I felt somewhat like a missionary but it was fun...

One special Wednesday evening we were anticipating viewing a long-awaited film starring Humphrey Bogart, but luck did not seem to be with me. An emergency Jeep accident came in shortly after six o'clock. The patient was a soldier...

(Dr. Albert Rood, 1973)

Dr. Albert Rood.

There were many doctors, dentists and nurses over the years. Perhaps one, Dr. Albert Rood, (pronounced "rude" although he was the kindest of men) really depicts how the Fort Stanton location not only helped him recover from asthma, but also fixed in his mind forever a special place that greatly influenced his life. His vivid descriptions of the Fort grounds and life style are valuable today for they tell a simple but elegant story of healing and adventure. His account also describes the life and times of Lincoln County, New Mexico during the World War II era and introduces us to many of the local characters he met while on duty at the Fort. His adventures in the Capitans, the Sacramentos, and along the Bonito valley document for us today an era long gone but obviously one that is remembered warmly and with much nostalgia.

The following are excerpts from a lengthy essay he wrote in 1973 with several anecdotes that depict the era, decades before when he refers to the fort as his *"Shangri La"*. This account is courtesy of Danna Henderson, and has been edited for length, but the words have not been changed.

Dr. Townsend standing in front of new hospital, c 1936.

MEDICAL MYSTERIES IN A LAND OF MANY CULTURES

By Albert C. Rood, M.D

"My Shangri La does not exist in the faraway Himalayas, as you might imagine, but in the Land of Enchantment located between the Sierra Blanca Mountains and the Capitan Mountains at an elevation near seven thousand feet. Old Fort Stanton was Paradise for me...

"...Only a country boy like myself would fully appreciate such a spot. The clear cool air, the beautiful western mesas, the fantastic starlit nights, the riding horses, the mountain trails, the cattle, the wild game, the trout fishing, victory gardening, the golf course, and many other forms of recreation available were all a joy I will never forget.

During the Second World War I requested a transfer to Fort Stanton because I was slowly choking to death from bronchial asthma in Mobile, Alabama. Of course I did not realize what a privilege it was to be stationed at this high altitude facility. Only doctors were sent there who had pulmonary disease, and so were not required to work very hard. My duties were usually finished for the day by four o'clock.

When I came to New Mexico for the first time, my arrival was by train in Carrizozo late one afternoon. I watched the landscape constantly all of the way from Dalhart, Texas, and was happy to see we were setting into mountainous country.

After a steak dinner at the café counter in the depot, a government car arrived to take us 35 miles to Fort Stanton. Winding through hills we came to the town of Capitan, past Titsworth's General Store—that carried just about everything—and then four miles to the Fort.

Previously I had advised the Commanding Officer of my arrival date and time. Then to my pleasant surprise I received a friendly reply scribbled on a piece of yellow tablet paper implying my new assignment might be more informal and relaxed than formerly experienced in the service. This thought of simplicity could not have been more true. Everyone was called by his first name, and little attention was given to protocol or rank. This situation was fortunate for me, because I was only a reserve medical officer of the lowest status.

The buildings at the Fort were located in an area surrounded by a large military reservation. There was one small two-story hospital of less than one hundred beds on one side of the parade grounds. On the opposite side was a long building constructed of thick limestone blocks which gave the impression it could be fortified to offer some resistance to Indian attack. That building contained several apartments. We lived on the ground floor at one end of the structure. Those were very comfortable quarters containing beautiful mahogany furniture. It was also interesting to learn our wing of apartments had had one formerly occupied by General John J. Pershing when he was a young lieutenant. On a third side of the parade was a small store and the administration building.

The more seriously ill patients were treated in the hospital, but those who had improved sufficiently were assigned to cottages located on a hillside in back of the administration headquarters.

Original nurses quarters.

Courtesy Fort Stanton 'Kids' Archives.

The cottage patients received a great deal of bed rest and fresh air, but at the same time were allowed mild activity. The cottages were heated with wood burning stoves—the firewood was mostly cedar, and in the evenings when I was officer of the day walking around the parade the aroma of the cedar burning as well as the night sky full of brilliant stars added to the magic of living. Such beautiful quiet nights I have never experienced elsewhere—and I should know for I was up half of the night delivering babies or doing emergency surgery for over thirty years.

Courtesy Fort Stanton 'Kids' Archives.

Nurses at Fort Stanton - c 1920.

Courtesy Fort Stanton 'Kids' Archives.

Christmastime in the Hospital Ward of the Old Army Barracks - c 1923.

Below - Patients enjoying Christmas dinner.

Courtesy Fort Stanton 'Kids' Archives.

In addition to the structures mentioned there were a number of attractive homes within walking distance of the parade. Warehouses, shops, a laundry, a central heating plant, two chapels, a meat plant, recreation buildings, barns, corrals, dairy buildings and even pig pens made up that self-contained facility. Some of the civilian workers lived in Capitan or Lincoln off of the reservation.

The area around the Fort included mesas covered with gramma grass, foothills, piñon trees, junipers, cedars and a few ponderosa pines. A number of wells with windmills were properly located for the thousand head of Herefords and the fifty head of quarter horses. Also, sixty Holstein milk cows and approximately two hundred hogs were maintained by the farm department.

I was more informed concerning the ranching activities than most of the other doctors because our next-door neighbor was Clarence Boyd, the farm superintendent, a graduate of Colorado State University, an ambitious, progressive, self-deprecating fellow whose company I enjoyed.

Courtesy Claire Boyd Fluharty

Dairy herd at Fort Stanton.

Bonito Creek (actually River) ran through the reservation not far from our quarters, a clear sparkling little stream well named, shaded in part by native black walnut trees. The creek provided irrigation water for the fields of alfalfa and our victory gardens.

Our Commanding Officer Dr. Kunkel was a trim, capable, well liked man with a twinkle in his eye who understood "the way of the West" and the

Courtesy Claire Boyd Fluharty

Coal wagons at Fort Stanton, c 1919

Courtesy Claire Boyd Fluharty

Hay truck deliveries to the Fort Stanton dairy, c 1920's.

feeling of its people. He was an arrested tuberculosis patient and had "chased the cure" in Tularosa himself a few years earlier. Unlike some COs, he had the good judgment to give employees as much responsibility as they could handle. He did not hold the reins too tightly, so to speak, so workers had a pride in their accomplishments and were stimulated to do better work by his suggestions.

My mind was filled with Public Health Service regulations pounded into me before coming to New Mexico so I was careful what I did. Often we had children or adults come to the hospital from various parts of Lincoln County without authorization to be treated in a government hospital. These cases were often emergencies with fractures or acute abdominal conditions. One day I asked Dr. Kunkel what I should do about one of the emergencies and he said with a twinkle in his eye, *"Al, you may do whatever you wish around here as long as you don't tell me about it."* His remark was pleasant and reassuring, but I knew he did not mean to be quite that liberal.

Courtesy Claire Boyd Fluharty

Hay wagons at Fort Stanton.

As far as social life was concerned, the name of the game was bridge, as was the case on many posts. The Wednesday night movies were always well attended, as I have mentioned previously, and the films were selected carefully for viewing by both children and adults. Youngsters came in large numbers and occupied the front third of the theater.

A club house was built near Bonito Creek away by itself where square dancing and ballroom dancing were enjoyed. Jerry had the inspiration to build the structure so we could be noisy and enjoy ourselves without disturbing any of the patients.

The area around the hospital and the parade grounds was overrun with dogs. The pet population was increasing so rapidly although… these dogs were strictly pets. As I recall, most of the dogs were friendly mixed breeds—and some form of birth control of these (animals) was lacking.

A majority of the community met with the commanding officer, discussed the issue, agreeing spaying the (females) would be an appropriate measure to be adopted. There was no veterinarian anywhere near, so the physicians were asked if they would help with the problem. This was not such an unusual request because all of us who had been through medical school had experience with dog surgery in physiology, pharmacology and surgical anatomy.

Dr. Iverson and I agreed to operate on the bitches, and Dr. Pollard volunteered to give the anesthetics. Pat Page, the surgical nurse, prepared a number of small, sterilized surgical packs containing instruments, sponges, towels and bandages needed for our canine patients. A Saturday afternoon was designated for the mass undertaking, which was not very popular with the children….

Our community contribution proved to be successful, and as I recall we did not suffer any direct criticism, but several of the children felt certain reservations about the mean doctors, and how merciless we had been to their dogs one weekend.

Serious efforts were made at victory gardening at a time when home gardening was especially encouraged during the Second World War. Families were assigned plots of ground near the creek. The area had been plowed and fertilized with liberal amounts of manure from the dairy corral which Clarence Boyd had supervised. I loved to garden, so ordered our seed from the Rocky Mountain Seed Company in Denver that promised to produce thriving high altitude plants. In 1943 our garden was a great success. We had such an overabundance of produce we canned corn, green beans, peas, tomatoes, made gallons of sauerkraut, dried many vegetables, and even furnished the hospital cafeteria with bushels of fresh head lettuce.

Original Publication: Washington, D.C. Agriculture Department. War Food Administration

1945 Victory Garden Poster
Artist: Morley, Original Size: 27"x19"

After work in the hospital about four o'clock in the afternoon, I would walk home, take off my uniform, put on my Levis, and go down to the creek bottom to work in the garden until dinner time. Usually there was an irrigation ditch full of water…

One Fall, we prepared twenty four cans of corn—when I had the *unorthodox idea* to wheel them, in a little wagon, all over to the hospital and pressure cook the entire number of cans at one time in the *surgical autoclave sterilizer*. Naturally the procedure saved much time, but Hope Kunkel spied what I was doing and made her husband do the same. The only problem there was Jerry went off to have a Coke leaving their corn in the autoclave too long, and it all turned a dark brown color. He was in deep trouble at home that day.

It wasn't many months before my health and strength improved and I found a friend—a lover of the outdoors, a real diamond hitch horseman—to go exploring over the hills on horseback with me. He was Slim Hendren, quartermaster of our post.

Photo by David L. Tremblay

Steps leading to the 9-hole golf tee.

After riding over the range, to the Fort Stanton Cave, and to Goat Springs, we… obtained a fishing permit from the Indian office of the Mescalero Apache reservation… we then rented a rather tricky old mule at Alto and packed our tent, grub and other camp gear on the beast. We began a six mile hike into the mountains near where the Ruidoso ski run is now located. Slim led our pack mule and I walked behind as a precaution, but even so halfway up the trail our pack animal jerked the halter rope away from Slim, quickly circled around me through the timber, and hightailed it for home. Fortunately, about a half mile back we had passed through a gate and were able to catch him and tighten his diamond hitch before starting uphill again.

The old Indian trail we used was too steep, had no drainage built into it at intervals, and consequently it was badly eroded. A deep gully had formed in place of the trail. When we finally arrived at our destination it was in a mountain park not far below the peak. A sparkling clear branch of the Ruidoso River rushed along the left side of the meadow, then continued precipitously down into a box canyon. Around us was thick, virgin timber. Most of the spruce trees along the edge of the clearing ranged up to four feet in diameter. Slim hobbled the mule, we quickly unpacked and set up our camp near the stream. A memorable three days followed. We both enjoyed the striking beauty of our surroundings, the quiet solitude, a drink of whiskey with snow water in a tin cup before supper in the evenings, the outdoor cooking—and we easily caught all of the trout we wanted to fry.

There was a nine hole golf course at Fort Stanton over by the East pasture only a few hundred yards from the parade grounds. One morning we were surprised to see a Navy fighter plane, completely intact, on the golf course. The pilot said he was from San Diego, he became lost and thought he was flying over Mexico but then saw our American flag and decided he better try to land. There was not enough room for the plane to take off from the golf course. So in a few days a large semi truck from a San Diego base hauled the dismantled plane away.

The Post had many stimulating, challenging activities to occupy my spare time. I built a large number of rabbit hutches with automatic feeders to be used by rabbits sent to us by my brother-in-law. Two huge pregnant Flemish White Giant does as well as a buck arrived from Oakland in excellent condition. I obtained a book entitled *THE CARE AND MANAGEMENT OF RABBITS* ...which served as a guide in my new adventure. Nevertheless it was not long before we had more rabbits to eat than we could enjoy, for even though the meat was white, tender and looked appetizing, we could not stand to eat rabbit once a week.

With all of the beef and pork raised at Fort Stanton, and the ham and bacon cured in the smoke house, we were most fortunate because all cuts of meat were just thirty five cents a pound. However, we could not overindulge because we had to have meat stamps. Restrictions were enforced as they were all over the country during the Second World War, but the Spanish (local Hispanics) men employed at the hospital as orderlies did not like much meat in their diet and gave me some of their stamps.

Bert Minter, a fence rider for the government and a typical cowhand, organized a deer hunt one fall in which I was included. We drove up in the Capitan Mountains before daylight one morning and Bert nailed his buck before the sun's rays were visible. None of the rest of us had even fired a single shot. After quickly dressing and hanging up his deer, Bert devoted himself to cooking and working around camp. I had a good time being out with an experienced, old time hunter. I remember he carried a little octagon barreled 25-35 Winchester carbine, a collector's item today. He only had simple iron sights, but he knew how to use the ancient rifle effectively. We had steak and hot biscuits prepared in camp style Dutch ovens with of course plenty of hot coffee, and fresh hot apple pie for dessert.

One of the forms of family entertainment in the evenings was the Wednesday night movie. Jack Shaw was our faithful, dedicated projectionist who gave our little community weekly films that never failed to draw a good crowd with many youngsters occupying the

Photo by David L. Tremblay

Typical prong-horned antelope on the Carrizozo range.

front seats. One special Wednesday evening we were anticipating viewing a long-awaited film starring Humphrey Bogart, but luck did not seem to be with me. An emergency Jeep accident came in shortly after six o'clock. The patient was a soldier, and when he took his cap off his scalp fell down over his face in broad strips like a partially peeled banana. With the help of a little novocaine we washed the sand out of his wound. A very large area of bare skull was plainly visualized, no evidence of a fracture discovered, so we quickly proceeded to baste up his lacerated scalp using fine silk sutures. Both Dr. Iverson and I sewed, then applied a pressure dressing—and made it to see Bogart almost on time. Somehow we got by with it, not violating too many principles of surgery, and the patient healed quickly without any signs of infection.

Today of course we would have x-rayed the skull, repeatedly irrigated the wound, carefully debrided the torn scalp, controlled the bleeding meticulously, and used very fine synthetic interrupted sutures spending well over two hours. I doubt if the results would have been any better. In those days we did not have to worry about *"defensive practice."*

There were many poor young families with babies in Lincoln County, especially in the towns of Capitan, Lincoln and San Patricio, who rarely had medical attention, so I organized a well baby clinic one day a week to try to improve the situation. I felt somewhat like a missionary but it was fun. We taught many of the young Spanish mothers how to prepare lactic acid formulas using evaporated milk. The method seemed practical at the time because none of them had ice boxes or refrigerators. At least it was gratifying to see a reduction in the number of infant diarrheas. We dispensed vitamins, did immunizations, kept records of weight gain, began feeding solid foods gradually, ordered blood counts, did tuberculin testing, etc.

During the Second World War, Dr. Turner in Carrizozo was the only physician in the county, and he had more than he could do even though he worked night and day. I learned to know Dr. Turner quite well, and on occasions would catch a ride on the milk truck early in the morning to Carrizozo to assist the busy practitioner with some of his more difficult surgical cases. In those days I did not even own a car. The Carrizozo hospital originally was a two story house converted to care for the sick, and we operated under an ordinary blue light bulb dangling from the ceiling. Once of the nurses gave open drop ether anesthetics for such operations as gall bladders, hysterectomies, ovarian cysts and even thyroidectomies on days I assisted. Dr. Turner was so busy, and gone on house calls so often, most of the time he was not around to deliver babies at the hospital, but the young mothers said they especially liked to go to Dr. Turner because his nurse did such nice deliveries, so it seemed to work out all right.

There were a number of complicated cases of tuberculosis that needed the attention of a chest surgeon. So it was necessary to secure the professional service of someone capable in that field. Dr. P.G. Cornish of Albuquerque was a pioneer in chest surgery in New Mexico who gave freely of his time, traveling to Fort Stanton nearly every month during the Second World War, but he was never adequately reimbursed for his efforts. He did not seem to mind, did not practice medicine for the money in it, and I am sure he felt his work at Fort Stanton was a contribution to the war effort.

I had heard of Dr. Cornish long before I arrived in New Mexico. His surgical judgment, experience, skill, speed and capability was almost legendary among United States Public Health Service Medical Officers who had previously been stationed in this Land of Enchantment.

Cornish was a surgeon who wasted no motions, and would often finish a difficult chest case in half an hour. I always assisted, and it was all I could do at first to keep out of his way. His speedy style of surgery is not considered important today, but then we were working on frail, far advanced tuberculosis patients, at an altitude of seven thousand feet where the atmosphere contained much less oxygen than at sea level. Also we lacked specially trained anesthesiologists like there are today.

My basic salary for my professional work was only one hundred and sixty six dollars a month, but we had so many fringe benefits it should not have been so distressing to me. Fringe benefits included well furnished quarters, utilities, steam heating and laundry. Milk and meat were very inexpensive, horseback riding and golf were available at no charge. We had more than enough social life with card parties and dances as well as movies once a week with an admission fee of only twenty five cents. Even a haircut only cost thirty five cents, and was done expertly by a convalescing Filipino. Furthermore, there was nowhere much to spend money except at Titsworth's General Store in Capitan, and lacking a car we did not go there very often.

One of my dear friends at Fort Stanton was the lonely, elderly French priest who had quarters in the little stone Catholic Chapel. Father Jules Moline spoke rapidly with a heavy French accent, he ate too well, and drank more than his share of red wine. The good Father came to me periodically to receive treatment for chronic gout. He loved to talk about the past, and indeed he had exciting memories to relate. Before training for the priesthood he had been a member of the French Foreign Legion in North Africa. He planned later to study medicine, but while he was a trooper in the Legion his battalion suffered a severe epidemic of typhoid fever, hundreds of his comrades succumbed, he survived but was left so horrified by the ordeal he gave up the idea of becoming a physician—and later went into the priesthood, coming to New Mexico during the nineteenth century when he had to ride horseback to make rounds in his assigned parish. Father Moline especially loved for me to bring him endive and romaine lettuce fresh from our garden so that he could prepare salads he remembered from his childhood in France.

Father Moline drove a good-looking Plymouth sedan, and not infrequently invited some of the nurses to go with him to Roswell. The girls appreciated his generosity, (and) needed to do some shopping occasionally...

One of the most unusual characters I knew was Bill Hale, chief cook for the hospital. He and his family lived in one of the new houses about four hundred yards east of the parade grounds.

Patients at Fort Stanton relaxing in hammocks, c 1920.

Courtesy Fort Stanton Kid's Archives.

For many years in the autumn before we arrived at Fort Stanton a rodeo took place in the area now occupied by the German prison camp. There was an abundance of excellent food prepared by Bill Hale including a big outdoor barbecue. Unfortunately this annual affair was discontinued and the rodeo grounds relocated along the highway to Capitan. It seemed nearly everyone recalled the good times they had at the old rodeos.

Photo by David L. Tremblay

Catholic Chapel – Fort Stanton, c 1930's

Rodeo Days at Fort Stanton
c 1930's

Courtesy Claire Boyd Fluharty.

Bill Hale and his two sons were renowned wild turkey hunters. They often hunted in the Capitan Mountains, traveled very light, not even setting up a tent or using sleeping bags. After supper was finished and their campfire coals had died down, the ashes were

scratched aside with a stick, the warm rocks moved out a few feet, filling the heated area with dry leaves, and then they just crawled into the leaves on the ground to sleep. No time was wasted washing up, quickly they became black and dirty, no doubt adding to their camouflage and in part increasing their success.

Wild turkeys possess extremely keen eyesight and if there is one thing that alarms them it is the sight of a clean, white human face shining through the pines. Therefore, many serious turkey hunters blacken their faces with charcoal and wear camouflaged clothing so they will be less conspicuous. Perhaps I washed too much, for I was never successful in bagging a wild turkey!

Eventually a letter arrived from Washington containing my discharge papers releasing me from further military obligations with three weeks of terminal leave. At last I was going to make plans about private practice. For years I had my heart set on having a solo

Courtesy Fort Stanton 'Kids' Archives.

Dr. Westhaven and Bill Hale on a hunting trip.

practice in a small town. Perhaps I would have stayed longer with the U.S. Marine Hospital Service if I could have been sure of being at Fort Stanton…"

And as he had arrived by train in Carrizozo, so he departed from Carrizozo several years later. No doubt Dr. Rood was leaving with regret, yet as a young physician he left with a feeling of excitement and anticipation.

"I felt I was on the threshold of a new life." he wrote.

And so he was.

He set up private practice in Albuquerque, New Mexico and was a successful and honored surgeon for another three decades.

Photo by David L. Tremblay

Photo by David L. Tremblay

Photo by David L. Tremblay

"Erbaut (Built) 1944, detail photo above main entrance to gymnasium.

Photo by Joseph Arcure

Chapter 5

The Winds of War
1939-1945

In ruins now, the gymnasium built in 1944 by the Germans stands silent in the wind.

(The German Internment Camp)

"I learned geography by helping my father trace the path of WWII on a large map he had on the wall and worrying about the relatives that so many of us children had who were caught up in the war."
(Danna Henderson, email communication, April 14, 2009)

December 19, Captain Wilhelm Daehne, on orders from Hitler, knew exactly what he had to do. For weeks he had been drilling his crew for this contingency... (Life Magazine, January 1, 1940)

"Nazi spy hunted since June falls into G-men's lap...a routine check on individual crew members brought Schluter to light..."
(The Times-Picayune, January 15, 1940)

"I distinctly recall what I believe were Nazi soldiers singing the Third Reich marching songs of war...we lived right across the river from them... and they sang well into the night keeping us awake..." (Tom Gould to author)

… "another Christmas Eve and much water has passed over the dam. I am still alive, which to me is something a few months ago I would not speculate on. The world is in chaos, we are at war…"

(From Diary of Carl Ness, PHS TB patient, December 24, 1941)

For several years a rumor of hidden Nazi artifacts being discovered after the war was over intrigued me and I was wondering was it simply a rumor? Who had discovered the items and under what circumstances? I soon found out…

(Author, Lynda Sánchez)

"Yes, I am writing a manuscript on the Japanese and Japanese Americans who were taken to Fort Stanton and then later deported to Japan…"

(Tetsuden Kashima, author of Judgment without Trial)

Courtesy Jim McBride

Guard tower at Fort Stanton Internment Camp, c 1942.

World War II interrupted everyone's life in the United States in one form or another, and so it was at Fort Stanton. Today, sprawled across the river from the main fort, ruins of a huge gym, swimming pool, smaller *casitas*, rock lined paths, an adobe guard house, laundry and reservoir provide an idea of how German and some Japanese detainees lived during World War II. Over 400 German detainees called this place home, and while most of them were not considered to be *"enemy"* combatants (Nazis), they had to be isolated from the rest of the community once the United States declared war on Germany in 1942.

The story of a CCC camp turned upside down and transformed into the United States' first German Internment camp is a fascinating one. In the last several years, through oral history interviews and the work of dedicated authors delving into the past like Jim McBride and Dr. Tetsuden Kashima, a much more detailed account of what really happened is coming to light. And with the special assistance of concerned individuals who either worked at the Fort or grew up during the Public Health Service years on site, we are obtaining a more complicated view of the so called detainees, both German and Japanese.

The majority were German seamen rescued from a distressed luxury liner caught up in the *"winds of war"*. Were they really just seamen, or were they Nazis? Did they taunt the youngsters from across the river just for the hell of it? Were they kind and helpful? Did they plot escape, even though this was not by any stretch of the imagination a hell hole of a POW camp like those throughout Europe and Russia? And what about the spy among them? How did they live? What was their ultimate fate?

And then there were the Japanese Americans. Why were they sent to this camp to be housed alongside the Germans? What was the Justice Department thinking? Did they even exist at all since some locals never knew of their being assigned to this area?

Many questions remain unanswered, but for the purpose of this chapter only a survey of what happened will be presented.

While we can not recount the history of World War II, we must try to understand the circumstances, or the totality of what was happening here. Hitler became Chancellor of Germany (early 1930's) in a depressed nation that was in a severe economic and cultural downturn. Thoughts of the greatness of the German people in the past under Charlemagne and other leaders must have been bitter pills to swallow as the Germany of that era was not yet over their present circumstances caused by losing World War I. When Hitler promised prosperity they listened. Unfortunately they ignored the destruction of their personal freedoms; and as the *"winds of war"* crept ever closer, they reveled in the newly found power and prestige of the Third Reich.

Courtesy Jim McBride

(Above and Opposite Page) Luxurious interiors of the S.S. Columbus.

It should be no surprise that the majority of the younger men were loyal to Hitler. Many were graduates of the "Hitler Youth" camps where they were thoroughly brain washed. Even if they or their parents disapproved of what was happening, to disagree or say no was a death sentence. After more than half a century we sometimes forget the evil that infiltrated Germany and Europe during those years. No doubt many young crewmen were anxious to get back to Germany and join the battle. We know from interviews that they sat around their radios cheering Nazi victories! (Gould, Henderson)

There was resentment too on the part of the children who lived on the other side of the Bonito River and of the adults who had family serving in the armed forces. Having the CCC camp selected to house the detainees and to later observe the wire fence and guard tower being constructed must have been a stressful situation for both the Germans and the employees of the Fort.

The German detainees were lead by Captain Wilhelm Daehne, a wise leader who spent many months trying to locate a place for his crew. They were not military men but a crew off a pleasure ship, the SS Columbus. Photos depict a beautifully handcrafted ship, similar in appearance to the ill-fated Titanic. Elegant dining rooms and carved stairways leading to a trim and classic upper deck created an elite tourist atmosphere.

Courtesy Jim McBride

Unfortunately, as the winds of war swept over Europe, Hitler ordered all ships to sail for home or be destroyed rather than have them fall into the hands of the enemy.

The crew of the Columbus tried to sail out of enemy waters, but delay caused them to dock near Vera Cruz, Mexico. Hoping to sneak back to Germany was their only option, as British destroyers and cruisers plied the warm waters of the Caribbean. For several years the Columbus, with all its elegance and courteous crew, had instilled a sense of national pride in the country. News reports came in weekly about her many adventures in the balmy Caribbean and the people smiled.

Courtesy Jim McBride

On December 14, 1939 the order finally came to leave Vera Cruz and head for Germany. The Columbus had been anchored offshore since September. They had disembarked most passengers in Cuba and were now awaiting a good time to sail. Unfortunately, the weather was clear, and the British destroyers soon tracked the ship, ordering it to stop. The Columbus was unarmed and could have easily given up

Courtesy "Badt Papers"

USS Tuscaloosa

Courtesy Jim McBride

without the terrible destruction that was to come. Orders from Hitler had been explicit. Do not let any ship fall into enemy hands. Captain Daehne's men had practiced for days just in case they were caught. Each sailor knew what was expected in order to burn and sink the *"pride of Germany"*. With the British destroyer bearing down upon them, one wonders what went through their minds as Captain Daehne signaled his men to scuttle the ship?

Destruction was precise and complete. Slowly the lifeboats were lowered full of saddened crew members. Only a few, including the Captain, were left on board. Huge billows of smoke darkened the sky. Eventually, all but two men were out in the ocean on life boats awaiting their fate. The missing crew members were never located.

It was said that at the last moment the men could not believe the ship's Captain was serious as he shouted out his last orders. But he was dead serious. The sixth largest vessel in the world, a gleaming symbol of Germany's determination to rebuild not only the merchant fleet, but also her pride and place amidst nations of the world, was gone on

Courtesy Jim McBride

Courtesy Jim McBride

(Above and Opposite) The tragic sinking of the S.S. Columbus.

December 19, 1939. It had sunk within hours! Now the dilemma was what to do with the 570 plus seamen. The British destroyer could not provide enough room on board for all men in the lifeboats.

It was the Cruiser USS Tuscaloosa that took them on board despite crowded conditions. The good ship Tuscaloosa, was preparing for *"neutrality patrol"*. That is unpleasant at its best but this time they were heading south. It gave the men an opportunity to enjoy sunshine, warm weather and calm seas. That was ever so much better than chasing icebergs in the northern Atlantic.

How the fate of these two ships and their Captains came together is a unique tale of the seas. The "Badt" papers came into the possession of Ms. Barbara Heiser, a special volunteer from Fort Stanton. It so happened that one of her neighbors, one Phil Schreiber, who had recently moved to New Mexico, was related to Captain Harry Badt. After several phone calls Captain Badt's daughter sent documents and photos. And thus we have the story of two strong leaders from different sides coming together in the Atlantic that cold December day in 1939. One Captain was Jewish, the other a staunch German. We do not really know how strongly Captain Daehne supported the Third Reich, but we do know that he followed Hitler's orders to the very end.

"JEWISH CAPTAIN SAVED A NAZI CREW"

Headlines declared this the "STORY OF A JEWISH CAPTAIN WHO SAVED A NAZI CREW".

Professionally, Captain Harry Badt had had a notable career. He graduated from the U.S. Naval Academy in 1904 and had served valiantly in numerous naval maneuvers of note. In 1938 he was given command of the Tuscaloosa. He never let his religious faith interfere with his work, and in that professional manner he also welcomed the distressed German

Photo by David L. Tremblay

Captain W. Daehne, commander of the S.S. Columbus.

seamen aboard after they dutifully and tragically sank their beloved and beautiful Columbus.

We had not yet declared war on Germany.

Coming into New York Harbor they disembarked at Ellis Island.

The Directors of the HAMBURG-AMERICAN LINE (North German Lloyd) sent Captain Badt the following letter:

> *"On behalf of the North German Lloyd, we desire to express our grateful appreciation of the splendid task performed by you and your crew in picking up and bringing safely to port the Captain, officers and crew of the ill-fated SS '**Columbus**'.*
>
> *"This humanitarian and workmanlike accomplishment is in keeping with the best traditions of the sea and of the United States Navy."*

(Letter December 27, 1939, Badt papers).

"NAZI SPY HUNTED SINCE JUNE FALLS INTO G-MEN'S LAP."

TIMES-PICAYUNE January 15, 1940

They extended greetings of the season and thanked them for the kindness and courtesy shown.

We have also learned that a major Nazi spy was a member of the crew and finally tracked down by the FBI. The spy was an elusive fellow and the United States had been looking for him for months. Karl Schluter posed as a steward and served as a transmission line by which prized American military secrets came into the possession of the German high command. A routine checkup on individual crew members brought Schluter to light. Once the ship landed and the seamen were processed through Ellis Island (more than 570 men), they were one less at the end after Schluter was taken to FBI headquarters for questioning. (The Badt Papers)

The next phase of their odyssey was to be shipped from Ellis Island to Angel Island in San Francisco Bay. Captain Daehne, in the meantime, had been working with Immigration officials for several months to get his men home. Angel Island was but one stop, they thought, on the way back to Germany. Several women and young cabin boys and older men were returned to Germany, but mostly the men of military age just waited. The plan had been to find passage for the men via Japanese ships.

Captain Harry Badt, Commander of the USS Tuscaloosa.

Angel Island was a pleasant enough place and the seamen awaited their fate. After fifteen long months from the time of their arrival at Ellis Island, Fort Stanton in isolated New Mexico, was that fate. Captain Daehne had visited Fort Stanton and agreed the abandoned CCC camp would be adequate for his men with some added construction and clean up. The men's arrival in the Wild West of their imaginations was really happening. They began to settle in to a rather easy existence (compared to other POW or Internment camps) at Fort Stanton on March 1941.

The camp was self sustaining and very disciplined. The men rose early, showered and shaved by 6:00. Breakfast in two shifts was at 7:00 and 9:00. They worked until noon and again ate in two shifts; the same for dinner upon their return to barracks after 5:00.

Hoping the war would be over soon Captain Daehne also realized he must keep his men physically and mentally fit. Thus, building projects popped up everywhere. In 1941 the U.S. was still a neutral nation and freedom to roam the hills and do more or less what they wished was allowed.

Claire Boyd Fluharty

However, Pearl Harbor put an end to their hopes... *"I remember Pearl Harbor...we had gone to cut our Christmas tree in the Capitan Mountains...we had a radio in the car and when we turned it on we heard the terrible news...that really meant big changes in all of our lives!"* (Claire Boyd Fluharty interview)

Our nation was now at war, and things changed dramatically for everyone on both sides of the Bonito River. Other German *"alien enemies"* were transferred into the camp. This created tension among the original group of seamen.

The war dragged on month after month. Once the wire fencing and guard towers were built in 1942 the men realized it was a grave situation. Some thought they would never see their homeland again.

Despite the bad news of being unable to return home, the German work ethic and desire for an attractive environment prevailed. Willett White, one of the Fort Stanton "kids" wrote: *"Their crew functioned with discipline and hard work as they had aboard their ship. They refurbished the old CCC camp, built a gym, a huge swimming pool, tailor shop and a large mess hall. Attractive fountains and rock-lined pathways connected the barracks and other buildings. The men built little huts or 'casitas' along the river where they raised rabbits, chickens and ducks. They planted piñons, hollyhocks and other flowers and grew vegetables."*

German internees at work building the swimming pool, c 1942.

Courtesy Jim McBride

Visitors to the camp over the years commented on what an attractive place it was and that it had been transformed from a sterile work camp into a livable environment.

The U. S. Border Patrol was responsible for the everyday running of the camp.

Fort Stanton doctors provided health care as per Dr. Albert Rood's vivid description below:

> *"Across the stream to the south was a German prisoner of war camp operated by the U.S. Border Patrol. It was one of my duties to care for the medical needs of the prisoners. About four hundred Nazis lived in the enclosure. These young Germans were in excellent physical shape and had little illness except for a few fractures caused during their enthusiastic soccer games on Sundays. However, just a couple of years before I arrived at Fort Stanton the prison camp experienced a tremendous epidemic of trichinosis. Over four hundred of the Germans were seriously ill. It was the largest epidemic of trichinosis on record.... The epidemic was caused by eating smoked sausage prepared in their own kitchens. Obviously the pork had not been sufficiently cooked. These parasites still caused some muscle pains in a few of the prisoners for years, and an occasional case of chronic phlebitis was treated as a result of that era of massive parasitic disease.*
>
> *"Surprisingly the camp contained two German doctors, one of whom escaped and was never apprehended. (There is no other mention of this except in Dr. Rood's account)*
>
> *"The prisoners kept busy during the warm months cultivating sixty acres of irrigated land where they raised their own corn, beans, squash, cabbage, potatoes and other vegetables which they stored, dried or froze for winter use. Work groups filling several army trucks left the prison compound early each morning during warm weather to work in their gardens--and on several occasions prisoners escaped. The Border Patrol officers employed expert Apache trackers to hunt down the escapees. The Germans must have seen a few Wild West movies*

Swimming pool and barracks at the German Internment camp.

Courtesy Jim McBride

at some time in their young lives, for they were deathly afraid of the Indians and offered no resistance when discovered, pleading not to be scalped.

"I had to match wits with the prisoners on occasions, for many of them were very homesick. They would describe in detail symptoms of various diseases in an effort to be repatriated to their homeland. One Nazi prisoner in the hospital for minor surgery even locked himself in the bathroom and slashed his wrists with a razor blade, doing a thorough job of cutting nearly all of his flexor tendons as well as the radial arteries—a most regrettable reaction…I have wondered since if he had conceived of my efforts as a form of torture he had received. By the time we had removed the hinge bolts and opened the bathroom door, he had blood everywhere. The bleeding was promptly controlled and he survived, but had to be transferred to a psychiatric hospital in Fort Worth for treatment of a severe depression."

SWIMMING POOL AND GYM

Though there were nine who tried to escape (some via tunnels out to the Bonito River), they were soon returned to the camp. The Germans were very ingenious in their escape technique. What to do with the dirt was always a problem. The men simply placed the dirt into garden plots that innocently kept taking on more and more dirt. The guards, at first, believed it was the men expanding the vegetable gardens. The escapees were easily caught because they had no idea of the vastness of the New Mexico landscape and they stuck out like a sore thumb. Jim McBride stated that the closest (escape) to success

Courtesy Jim McBride

Connecting walkways constructed by Germans at the camp.

was in November of 1942 when four men were gone for three days. One of the four was shot in the leg by a local rancher so that ended their escapade.

Generally, a good working relationship was created with area communities and local ranchers used them as day laborers. Johnny Thomas was one of those ranchers and he said they were good workers and loved to talk. Germans were enamored of the history about the Old West. They were delighted by tales of Billy the Kid and Apache warriors. In fact, several Germans returned after the war to show their families where they had been confined. Some even became U.S. citizens.

Perhaps an artifact found by Jeannie Adams, a volunteer with the 2004 archaeology survey team, sums up this history. A medallion had survived the elements along the side of a steep arroyo in back of the camp for several decades. One more storm could have covered it for yet another 50 years. In German it says: San Francisco, *Sangerfest*, 1939. What a story it could tell! We know they were at Angel Island near San Francisco before they traveled via train to Carrizozo and then bused to the Fort. How it came to be in this spot no one knows. Possibly one of the men out hiking had lost it on the downward trek from the top of Bullet Hill.

Detail of Angel Island Medallion c 1939, found by Archaeology survey crew in 2004.

Photo by Joseph Arcure.

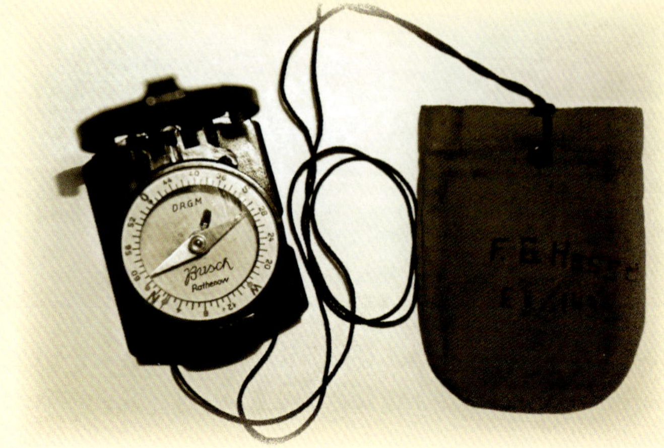

Photo by David L. Tremblay

Yet another artifact, was one given to the author by Roberta Haldane. Ms Haldane is a descendent of area pioneers, a generous and special lady and member of the Lincoln County Historical Society. She sent the author a small box with a donation for the Fort Stanton Museum. A plain canvas bag contained a very old Busch compass owned by one F. G. Hesse. At first it was thought to have belonged to an infantryman stationed at the Fort in the 1870's. But the composition of the compass being of a hard element known as Bakelite pointed to the WW I or II era and the name, being German, yielded yet another fascinating possibility and personal story. We now believe that it belonged to one of the German POW's. All of the Nazi soldiers wore compasses around their necks-a matter of practical survival. Somehow this one was never seized by the guards and found its way to Ms. Haldane.

Looking at the material from the hand sewn brown canvas bag makes one think of the brown shirted Afrika Korps Germans who may have been imprisoned at the camp toward the end of the war. Did it belong to one of them? That is the most plausible explanation as to its origin.

Both accounts are fascinating. When the new and enlarged Museum opens at the fort, Ms. Haldane's compass will hold a prominent place for all to behold.

When interviewed about their memories of the Germans and Japanese, the majority of the children who grew up at the fort noted that they did not like them (the Germans), that they were cruel and taunted the children and were quite aggressive. In retrospect it is possible that the children took on the attitude of their parents because we were at war with the Axis powers. Nevertheless specific comments do stand out. For example:

"The only things I really remember about the Germans are them taunting us kids with Cokes and candy bars when we couldn't get them during the war, and my mother being locked in the camp overnight treating the trichinosis patients. My parents had nothing good to say about them. But I was seven when the war started and eleven when it ended, and although my father followed its progress in Europe every night it was all pretty distant to a kid my age… the Fort Stanton store stocked ersatz candy bars that tasted like saw dust and had carob coating…they were awful and we kids hated them. We all were very envious of the Germans because they had real candy bars, Cokes and a swimming pool."
(Danna Henderson and several others tell the same story)

"With regard to the question about if I felt that all the Germans were bad - at the time 'yes'. I am sure that actually only a few were "bad", but we were at war with the Germans and it was a different time. Those few "bad apples" were cruel enough that they enjoyed baiting small children by pouring Coca Colas on the ground and stomping chocolate bars and oranges which they knew we couldn't have. I think we would all agree that type of personality was both cruel and disgusting. Unfortunately a few bad apples were able to poison the attitude we had for the entire camp.

"My Mother was a nurse's aide and was helping one of the Doctors work on a prisoner who said something in German. The Doctor fired back in German and the prisoner got very pale and quiet. Later he told Mom that the guy had made a nasty remark about the doctor being Jewish. When Mom asked him what he had replied, the doctor said he told the German that if he said anything more he was going to modify his privates during the scheduled surgery. Mom said after that the German was VERY cooperative.

"My Mother would sometimes come home depressed and in tears after working with our sailors who had been torpedoed. She said some were in very bad shape emotionally and that in turn affected how we felt about the Germans." (Tom Gould)

"Louis Padilla, an old classmate in Capitan High School called me as a result of my letter. His father was a guard at Ft Stanton the whole time. He went there a lot, and was given many things by the German—including candy which he said came from their Red Cross—everything was written in German script and in German. When they returned to Germany, they couldn't take so much stuff with them, so they gave a lot away..." (Claire Fluharty)

The author's interest was piqued when in 2004 and 2005 she was part of an archaeological field survey team funded by the National Park Service and the Bureau of Land Management. They spent several days on site not only at the fort, but the Internment camp as well.

What triggered her interest most was the rumor of hidden Nazi artifacts being discovered in an underground room after the war was over. Was it just a "rumor"? As rumor so often does, it kept circulating and yet no one knew the origin. Finally, after more than three years, Mr. Tom Gould was located living in Texas and asked about his adventures in Fort Stanton Cave. (see cave chapter) Several emails later when the rumor was mentioned, Tom revealed to the author that:

"I was among the group of kids who found this material. It was in an underground room beneath the mess hall. We found the location after we were allowed to explore the camp once the Germans were gone. Another boy and I located a hidden doorway behind some garbage barrels. As boys will be boys we pried open the door. The room had flags, pictures of Hitler etc. We were so mad we did burn the stuff; however, we also reported it to Ray Baker, our Scoutmaster. He inspected it before we burned it.

"Also we found more near the fence on the river side where they had several small huts with duck ponds. They had used the dirt from an escape tunnel to build the gardens and huts. We kids explored these huts and the one that attracted my interest was unique where the owner had plastered bottles into the wall allowing light into the hut. While examining the place I found another hideaway with the Nazi flag and a picture of Hitler. We kids made short work of that contaminated place! Obviously to me now, as then, a lot of those guys were hard core…"

(Tom Gould, interviews 2007 and 2008)

Possibly in the beginning when the merchant seamen came to the camp with Captain Daehne and began most of the clean up and building they were not so hard core. Later, however, other Germans were transferred in and toward the end of the War a special segregated camp was constructed outside the original compound. It was probably from that compound that Tom Gould had listened night after night to German marching songs.

Photo by Joseph Arcure 2004

Archaeology Field Survey Crew on site, surveying part of the Internment Camp.
(left to right) Bob Davis, Lynda Sanchez, Louise Ritch, Dorothy Foyt, Jeannie Adams, Charles Haecker, Jim McBride.

Courtesy Dana Henderson / Matsumoto

Japanese detainees held in Santa Fe, New Mexico. Some were transferred to Segregation Camp No. 1 (Fort Stanton).

THE JAPANESE AT FORT STANTON:

Thus, we have a mixed bag of opinions, experiences and emotions with regard to the detainees. Accounts of Japanese or Japanese Americans interned at the camp reveals yet another story of frustration, wrongful imprisonment and lives disrupted by war. After more than 5 decades of silence, research in recent years pertaining to camps for Japanese Americans and especially those interned or detained right after the bombing of Pearl Harbor, is out in the open. There were dozens of camps throughout the United States. Some were for Japanese American families who needed protection from an angry U.S. populace. Baca Canyon (Raton Ranch) up Salazaar Canyon was one. Others were for those suspected of harboring Japanese or supporting Japan in its quest to conquer the world. This is why the DOJ 'Segregant' camps were established. One was built at Fort Stanton within the area of the German Internment camp but outside the main area. The devastating bombing of Pearl Harbor destroyed faith and friendship and it has taken decades to rebuild some of those bridges.

Courtesy

Dr. Tetsuden Kashima, only a small child at the time of World War II, remembered a suitcase that was always packed and placed in the living room of his home. Kashima and his family were eventually sent to an incarceration camp in Utah and suffered greatly. He never understood why until years later when he began to unravel what happened to his own family during the war. During his many months of research in archives and libraries he discovered a very special story that could apply to the more than 100,000 Japanese Americans who had been interned during the war. It was in many ways his own story as well.

As a result of his valuable research, information leading eventually to Fort Stanton has revealed another chapter in the history of the main German Internment camp. We are grateful to scholars like Dr. Kashima and look forward to additional material in the future clarifying the mystery; the trials and tribulations of the Japanese families or single men who ended up in Lincoln County.

In a search for answers Dr. Kashima received a federal grant from the Civil Liberties Public Education Fund. He spent endless hours sifting through archives in New Mexico and Washington, D.C., uncovering intriguing information about these camps. In many presentations and in his book, *Judgment without Trial,* Dr. Kashima, Associate Professor of American Ethnic Studies (University of Washington), uncovered information about a Department of Justice (DOJ) camp at Fort Stanton, New Mexico, that held German and Japanese internees identified as 'troublemakers.' *"No one really knew about this camp,"* says Kashima. *"The government did not want people to know it existed."*

Located within the larger camp that held captured German seamen, the facility was called a "Segregation camp", and its 48 inhabitants (there are some reports of 58) were called "segregants." Kashima believes the terminology is significant.

> *"At the camps where some 116,000 Japanese Americans were held, they were called 'relocatees' or 'evacuees,'" explains Kashima, "with the implication that we evacuate people--in flood or famine situations, for example--for their own good. But the people at Fort Stanton were called 'segregants' and they were viewed differently."*
>
> (2007 speech to Civil Liberties Symposium II, Twin Falls, Idaho)

Who was sent to Fort Stanton? Of the 17 segregants of Japanese ancestry, 10 were born in America but faced an unusual situation when, in 1944, the U.S. allowed Americans to renounce their citizenship in time of war, on American soil. *"Anyone who renounced his or her citizenship was treated as an alien," says Kashima, "and many were scheduled for deportation to Japan."* The others were not U.S. Citizens. First however, *"these individuals were brought to the DOJ camp at Fort Stanton." (Kashima) The total of the first seventeen were considered to be*

Courtesy Fort Stanton 'Kids' Archives.

Baca Canyon Camp where Japanese families were detained in 1942.

the most troublesome. Another report claims that the DOJ "also established a disciplinary camp for those termed "incorrigible agitators". Fifty-eight Japanese were said to have been sent there." (nps.gov/archive) (No time line could be found for all 58 as to when they came or departed.)

Border Patrol agents rather than military police were used to guard the Japanese as they also guarded the German detainees. Fort Stanton was a place where the Justice Department decided they would not tell relatives the location of the camp. They just called it Japanese Segregation Camp #1. All mail was supposed to go to Santa Fe, New Mexico and then routed back to the Fort Stanton camp. However, some of the relatives found out that their fathers and oldest sons were held at Fort Stanton. (www.artsci.washington.edu/news, 1998)

Other families were sent to Baca Canyon near Capitan for protection from angry non-Japanese citizens. They had come from Clovis. There were 10 men, 5 women married to these men and 17 children. The men had jobs working on the Santa Fe railroad, and after Pearl Harbor there was a lot of anger toward Japanese people. So, they were thought to be at risk and were taken to the camp. Eventually, in about a year, it was disbanded and the people were relocated to camps elsewhere in the US. (Information, courtesy of Claire Fluharty via email. She is also very interested in this topic and has researched much of the Japanese story.)

The Baca Canyon camp housed several families including the Border Patrol agents and their families who also lived within the compound. That was more of a place for refuge from mob violence than a detention camp according to some. In case of serious illness those detainees would be transferred to the Marine hospital at Fort Stanton. In a report written on May 25, 1942 the camp was visited by the U.S. State Department and the Spanish Consul. A description noted *"the camp itself does not resemble a detention station but rather a small country community."* (May 25, 1942 U.S. State Department report)

While many of the local residents of the county recall in great detail the Germans and their internment at the Fort, little has been written about their recollections of the Japanese Americans. Recently, Claire Boyd Fluharty has taken time to speak to some of the old timers and the following is based upon her research and interviews with Alfred Sánchez of Lincoln, Louis Padilla, Emmett Womack and others whose parents worked at the camp. Their comments are also included below. There are some discrepancies and repetition, however just like with the German internees, each individual had specific experiences that influenced their lives and thoughts about this period of time and are quoted as such.

> *"In general the memories are blurred but most recall Japanese being there from 1942-1945, probably on an intermittent basis until they could be transferred to specific camps elsewhere that had a total population of Japanese. The Germans did not care for them so the two camps were separated, one being upriver from the main German camp.*
>
> *"Emmett (Womack) said he worked on building some housing in the German Camp during a summer. He said it was for Japanese who had been sent there. He said the Germans didn't "like" the Japanese.*
>
> *"Anyway, that is the area where the seventeen "segregants", as labeled by Kashima, were put. I also believe those are the ones that terrified Ben Leslie and presumably other guards but, the Japanese had their housing built by the government. Lloyd Hulbert was the lead carpenter and he thinks it was about 1944."*
>
> (Claire Boyd Fluharty, in correspondence with author, 2008)

"According to Jean Sánchez, guards were afraid of the Japanese prisoners. They were put in a different area and had much higher walls etc.

"There was also a different group of Japanese that Alfred (Sánchez) knew and played ping pong with, and they were not in a heavily guarded site. Alfred brought food and other items for them from the Fort Stanton Store. He said that they were typical California Japanese Americans and somehow they ended up at Ft Stanton although not for long.

"But he also indicated that some Japanese were there early in the war—in 1942—and they were with the Germans at first. He said those Japanese had time to play with the kids—but the Germans, with whom they were mixed then, had no time. They were always 'too busy' to play. Alfred said he was about 12, which is a perfect fit for 1942." (Claire Boyd Fluharty)

Willett White verified the fact that Japanese children were enrolled at schools in Capitan and the Fort. He wrote: *"For a while the men and women with children lived at Baca Camp site up Salazaar Canyon. There were several children enrolled in Capitan High School. Two girls were enrolled at the elementary level at Fort Stanton supposedly because they were not accepted in Capitan. But they only stayed a brief time. One can just imagine their confusion as they were switched from one isolated area to another."*

When Claire asked others about any Japanese in school, several stated that they went to Capitan High School with "a bunch" of Japanese kids. They came from the CCC camp in Salazar Canyon, the place called Baca Canyon Camp. There were both boys and girls and they eventually left the school once they had their own.

Claire continued… "When that friend of ours died last December and we went to the funeral, I heard story after story of being in the camps. Most of the men picked up right after Pearl Harbor eventually served in our military as well. Our friend was an of-

ficial military translator in Tokyo after Macarthur went in and he stayed a few years after 1945 in the military and as a teacher. All of that was such a tragedy...."

June 2, 2008 correspondence, Claire B. Fluharty.)

As these comments and those below are woven together, the research done by Dr. Kashima ties things together and it is now clear that there were two types of detainees. At the very end of the war several also arrived for "repatriation" and those were probably the ones encountered by Tom Gould:

"When the Germans left and the fence was taken down there were some Japanese based there. In front of the gymnasium there was a fish pond. There was an old Japanese gentleman (old to me at the time) who was making lamps and other stuff out of cactus. This kind gentleman patiently showed me how to soak the cactus limbs and then clean and dry them to make lamps and other things. I don't have any idea how many were there, but the ones I met were all very nice and spoke English. The only negative thing I remember about this period was there was talk about a scandalous affair between a guard and one of the Japanese ladies, but it was very hush hush and we kids never knew details."

Dr. Kashima finds that people are often surprised to learn that individuals of German and Italian ancestry were also interned (in the United States) during the war. In fact, more than 11,000 Germans--mostly German nationals--and more than 2,700 Italian nationals, half of whom were Italian seamen, found themselves in various DOJ camps. That's in addition to the more than 17,400 Japanese Americans who were interned by the DOJ and the U.S. Army.

"This is therefore not just a Japanese American story." He also recently wrote an article for The Washington Post on the redress issue of Japanese Americans. *"There is still much to be learned about this subject,"* Kashima says. *"Vital aspects of the experience have yet to be explored. The story of the incarceration—and efforts to obtain redress—is a supremely American tale. As such, it should not and must not be forgotten."*

As the war drew to a close there were hundreds of happy men returning to their native land. Captain Daehne was paroled and returned to his home in January of 1945. Of the original 410 men who arrived in 1941, only 239 had spent the entire war at Fort Stanton. Four remained permanently as they were buried in a lonely and separate corner of the Marine Cemetery. Most were gone by July of 1945 and the German Internment camp closed officially in October of that year. One wonders what they faced upon returning to a bombed and devastated Germany?

It is, however, quite amazing to realize that years later many of them, with family in tow, visited the United States to show them where they stayed in the wild western hinterlands of New Mexico. These trips were made as tourists with a sincere desire for friendship and not anger or resentment.

When Jim McBride began his long journey to obtain information he made numerous friends on his research trips abroad, thus allowing him to fill in the blank pages of history. Dr. Kashima, too, has helped with additional details as have the several "children" or "kids" at the Fort in their often colorful and poignant stories. While there may be discrepancies, we are much closer now to finding out how the camp operated, and what really transpired. Mostly, the experiences were reasonably good for the detainees who were, unfortunately, victims of war.

After the war the staff and families with children benefited from much that was left behind. Some of the units were turned into housing, or auctioned off. The Border Patrol and Immigration Service turned over the camp and what remained to the U.S. Public Health Service. There were no more raucous marching songs being sung in German, and the Japanese Americans had also returned to their homes.

> Tom Gould wrote: *"Lynda you are correct in that we used those facilities a lot. On hot summer days the swimming pool was*

always a hub of activity. I know I must have spent hundreds of hours playing basketball in the gym. We also had parties; roller skated and generally used the gym a lot. In fact, I can't remember any periods of inactivity at the gym.

"Once after the big Capitan fire (the one where Smokey Bear was recovered as a badly burned cub) they covered the gym floor with freshly mowed gramma grass. When it dried we kids would run back and forth over the grass which dislodged the seeds. They then picked up the grass for the dairy cattle, recovered the seed, mixed it with some clay type fertilizer and molded it into pellets. The pellets were spread over the huge burn area by aircraft. The theory was that the pellets would stay intact and protect the seeds until there was a big rain. They would then dissolve and the seeds were planted and fertilized."

Perhaps that was a good way to begin the next decade, with rebirth and planting of seeds to bring the earth back to life. The scars from that terrible war, like the scars of the Capitan fire, have almost disappeared into the mists of history. The "winds of war" have disappeared and peace has put down deep roots. The winds now only gently blow the cottonwoods along the Bonito and beckon us to revisit history with open hearts and minds.

Cottonwoods along the Bonito

Photo by David L. Tremblay

SOURCES AND RECOMMENDED READINGS

Interviews or personal communication:

Claire Boyd Fluharty, many interviews and emails, 2007-2009

Gould, Tom (Skipper) many interviews and emails 2007-2009

Haldane, Roberta, personal communication February 1, 2004

Henderson, Danna Kuscianovich, many interviews and emails 2007-2009

Kashima, Tetsuden, personal interview by telephone, summer, 2006; emails 2009

McBride, Jim, emails and interviews 2001 to present.

Books, Magazines and other sources:

The Harry A. Badt papers, courtesy of his daughter, Esther, presented to Fort Stanton Foundation, 2008. Also on file with the Lincoln County Historical Society and the Lincoln State Monument office.

Life Magazine, January 1, 1940

McBride, James J., Interned, (Internment of the SS Columbus Crew at Fort Stanton, New Mexico, 1941-1945, Santa Fe, 2003 (self-published)

McBride, James J., A Photo Essay History Fort Stanton Marine Hospital, Public Health Service 1899-1953, Santa Fe, 2005 (self-published)

Kashima, Tetsuden Dr., Judgment without Trial (Japanese Imprisonment during World War II), Seattle, London, University of Washington Press, 2003

Courtesy NM State Monuments

Photo of CCC Camp c1940.

Photo by David L. Tremblay

(CCKW) Vintage WWII Truck, restored by Larry Holt of Capitan, New Mexico.
Photographed at the Fort Stanton Memorial Day Service, Merchant Marine Cemetery.

Kashima, Tetsuden Dr., in a speech before the Civil Liberties Symposium II at the College of Southern Idaho, Twin Falls, June, 2007

Rood, Albert, Dr., Personal Essay, "Medical Mysteries in a Land of Many Cultures", (unpublished) 1973 (see Chapter Four for the major content of this unique personal glimpse into life during World War II at Fort Stanton)

U.S. State Department report, "Report of the Raton Ranch, Civilian Detention Station near Capitan, New Mexico", (May 25, 1942) Report on visit to Baca Canyon camp.

"On the web"

www.Idahoptv.org (June 2007)

www.artsci.washington.edu/news (1998)

dic.academic.ru/dic

nps.gov/archive

newworldencyclopedia.org/entry/Japanese_American_Internees

bookmice.net/darkchilde/japan/raton

Photo by David L. Tremblay

Early cave explorers.

Chapter 6

The World Beneath Our Feet

(Fort Stanton Cave and Snowy River Passage)

"Shore and tis the Gates o' Hell." **Another:** ***"Nein, das ist der Backbone oaf zee Teufel."*** **(Devil's Backbone, an apt name in an appropriate place.)** ***"Knock it off and keep your eyes and ears open!"*** **hisses the corporal. They continue the scout.**

(BLM Cave Specialist Mike Bilbo's description of possible first entry into the Fort Stanton Cave, March, 1855)

Within only a few miles of the Ruidoso area lies a world, that while "beneath our feet... is an entirely unseen realm of great beauty, fragile biology, exquisite minerals and a window inside the very skin of Mother Earth... It is composed of... many caves ...that differ so much from the overlying surface environment that people often feel they might as well be on another planet."

(Dr. Penelope J. Boston, Cave Studies Director at New Mexico Institute of Mining and Technology (Socorro) in testimony before U.S. Senate.)

(Opposite) The Fort Stanton Cave Study Group, heading into the Fort Stanton Cave and Snowy River Passage, c 2009.

Courtesy Tom Gould

Adventuresome young cavers, 1955, tired, dirty and joyous.

Another fascinating aspect of Fort Stanton is the incredible natural beauty of the area and the rich flora and fauna as well as the geological formations both up top of the earth and those within the numerous passages of Fort Stanton Cave. Exploration of the Cave by troopers (Dragoons) began shortly before the fort was established and continues today with a dedicated group conducting surveys, creating scientific mapping and documenting of the cave biota and unusual passages or formations.

The two quotes above are worlds apart and are separated by more than 150 years. Visitors to the cave began from the moment it was discovered by Apaches and later the "locals". Then adventuresome types came for a "look see" and returned with tales of sights that were wonderful to behold.

All of this has come full circle and both the historical and biological/ hydrological research continues. Each year reveals more information about this fascinating world beneath our feet. The Bureau of Land Management will soon be overseeing a National Conservation Cave Area that will provide additional protection for this awesome and special place.

And, because we are usually unaware of this world, out of sight can truly be out of mind. On occasion we may have visited a place like Carlsbad Caverns or Kentucky's Mammoth Cave that are geared to tourist visitation. However, there are innumerable areas nation-wide containing unknown treasures. Cave researchers and volunteers have for years lead the fight to protect, interpret and educate the public about these hidden masterpieces and are leading the way here in New Mexico as well. Cave environments are fragile and just a crust from a sandwich, or one's breath can damage bacteria or other formations. That is one reason many caves are closed to the public at large and why a special permitting system has been developed.

FORT STANTON CAVE THROUGHOUT TIME:

Throughout the decades, locals visited Fort Stanton Cave attempting to explore it by using candles and lanterns. Dragoons and other military mapping crews explored and tried to get a handle on the winding passages beneath the terrain upon which Fort Stanton had been constructed. Centuries before the native peoples had walked along some of the corridors with smoking piñon torches in hand and Boy Scouts in the 1950's and 60's explored the cave with flashlights, a sack lunch, and no safety equipment, according to many old timers who were young boys at that time. Sometimes their parents just left them off outside the cave entrance!

So, let us step back in time for the moment and view the cave from an historical perspective...

Even the famed Lily Klasner wrote in her intriguing autobiography, *My Girlhood Among Outlaws*, about a "rumor" well known but not discussed on the frontier during the bloody Lincoln County War era. Apparently the Murphy/ Dolan gang was said to have used "the cave" as a hideout for the making of bootleg liquor. She remembered this well because her family's grist mill at Hondo did not grind the corn that came in heavy loads from the Fort Stanton/Lincoln area. Rather her father "cracked" the corn and sent it back to Murphy's in huge, lumbering wagons. Everyone knew that "cracked" corn was used in distilling of liquor. In addition, "Bootlegging was not the

Courtesy Lynda Sánchez Collection

Lily Klasner.

only contraband activity allegedly carried on in the cave…" counterfeiting was another! (Klasner, p 97-98)

Was this the Fort Stanton Cave of today? Lily Klasner and many others seemed to think so and feared going near the site because of the illegal activities. Others do not agree, but acknowledge that there may be other cave-like openings in the vicinity. Geologically Lincoln County has the kind of land forms where innumerable cave openings, Karst formations and other features indicate a major underground world to explore.

Soldiers from nearby Fort Stanton also made forays into the cave area from the mid 1850's.

THE FIRST DRAGOONS AND FORT STANTON CAVE

Mike Bilbo, Cave Specialist and Recreation planner for the Bureau of Land Management has done extensive research as well as exploration of Fort Stanton Cave. In his excellent paper on the First Dragoons he describes the following ventures into this underground world during the first decade of the Fort's history as follows:

> *"In 1855, a patrol of the 1st Dragoons from Fort Stanton, New Mexico Territory, explore a large limestone cave located about one mile north of the fort. Their horses tied up and under guard, the men slowly and carefully make their way down the steep, loose entry sink talus. At the drip line the musty smell of the cave assails them. The soldiers are dressed in the military clothing typical of the period: white wool shirts under dark blue wool shell jackets, sky blue wool kersey trousers supported by cotton galluses, leather boots, and non-descript campaign hats …*
>
> *"Down they descend into a dark, dank cavern for the first time - the first white men to explore this cave. Down into the gloom, carrying bulky whale oil lamps, ropes, haversacks, tin canteens,*

their heavy .44 caliber pistols belted around their waists and maybe carrying their musketoons, too. These men are young but they're veterans - tough 'ombres all right - they've been in some fights with Apaches, Comanches and Comancheros. They take it slow and cautious. Their lantern lights - their only outside reminder - flicker dimly on the walls, casting grotesque shadows all about.

"Following a main passage south and east, the patrol treads first through mud and water, and then up onto massive piles of limestone blocks covered by white, powdery rubble - one of them mutters, 'Shore and tis the Gates o' Hell.' Another: 'Nein, das ist der Backbone oaf zee Teufel.' (Devil's Backbone, an apt name in an appropriate place.) 'Knock it off and keep your eyes and ears open!' hisses the corporal. They continue the scout.

"Caving is a part of their duty - they must understand all aspects of the topography they are to patrol in the coming years. Fort Stanton has been established to protect regional settlers from Plains Comanche and the nearby Mescalero Apache.
"After some 2,100 feet of slow, careful progress and a slippery climb up a steep 20-foot mud slope, they are suddenly faced with a choice: the main passage bears away north and east, while to their

Dragoon reenactors inside Fort Stanton cave.

Courtesy BLM/Mike Bilbo

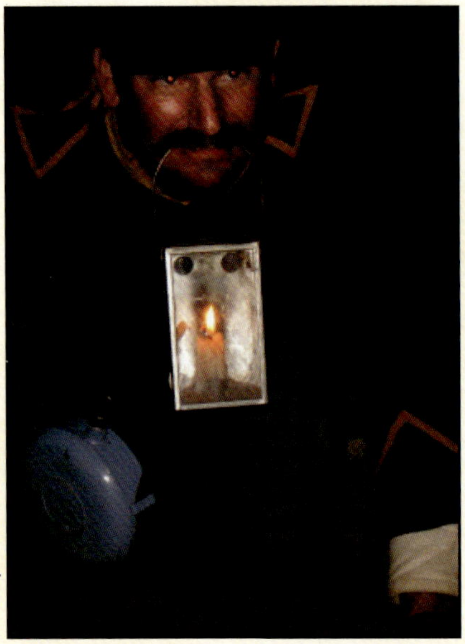

Reenactor Larry Pope carrying oil lantern as the Dragoons did in 1855.

right it pinches down into a crawl way. Being adventurous troopers, this intrepid band of recent recruits from Governor's Isle, which includes some immigrant Germans and a Scot, choose the crawl way. On hands and knees, carrying their lanterns by the bales between their teeth and dragging their gear behind them, they enter a broad, low passage and behold a most amazing sight: like a carpet of grass, thousands and thousands of crystal clear gypsum needles, each almost a foot high, fill the passage.

"The Dragoons push forward, but out of respect for this wonder of nature, they keep their passage shoulder-width only. This is not new to the Germans - they caved many times in Bavaria only a few years before. 600 feet later the party exits the crawl and stands up, 'Let's take five for a lit.' The soldiers have just come out of Crystal Crawl and are relaxing at the beginning of Decoration Passage."

(Courtesy Mike Bilbo, *"The First Dragoons and Fort Stanton Cave"* from Preface)

Examples of crystal formation inside the Fort Stanton Cave.

Early Non-Military Explorations and Descriptions about Fort Stanton Cave:

The following material came from an 1872 letter found in the Eve Ball papers. Written by Ash Upson a contemporary of Pat Garrett, (even helping with Garrett's account of the Lincoln County War), this reporter was sometimes referred to as a blow hard. Nevertheless he was a good writer and one of Lincoln County's first "intellectuals". Eve had it labeled "Fort Stanton Cave material" and entitled it: *ASH UPSON ON THE FORT STANTON CAVE.* His descriptions are vivid and colorful and definitely part of the cave folk lore and history. Below are a few excerpts from his adventure inside Fort Stanton Cave. Upson was staying at the Casey ranch in Picacho, about 20 miles from Fort Stanton:

"There is a cave about 20 miles from here and some 3 miles below the Fort, near the Rio Bonito, the extent of which has never been ascertained. I was in there for more than seven hours, traveling constantly, but could not in the multi-farious windings and spurs of the labyrinth, find a single termination of a passage, nor a passage where I was forced to return by the same one that I entered. In some portions of the cave, the roof is very lofty; in others it is so low as to compel one to stoop. There are beautiful ever-living springs scattered through the cave, and the stalactites with which the ceiling is hung would almost enrapture the senses of one who has never witnessed a similar scene…

"By the dim light of our candles, they blazed with splendor—their grotesque, grand, quaint, or ludicrous contour intensely discernible by the preternatural light, reflected from them… In one place a lofty dome seems to pierce the roof of the outer world, and in the center of the dome arises a lofty pyramid formed of crystallized rock and jasper. Down the sides of this pyramid drips bright, sparkling water, with a solemn monotonous sound, as though weeping at the presumptuous impotency of man. There are chambers that outshine the glories of the Alhambra….

"It confuses the sense to gaze so long upon the brilliant beauties of this place...General Carleton and several other officers were more than two days and nights in the cave. They burned a box of candles but failed to find a terminus."

Upson goes on to tell about a Mexican girl and her father who live by the cave. The girl gives tours to parties, and she also tells about Apache murder, and starving a group of Apaches inside the cave.

When asked about Upson's colorful descriptions, John Corcoran in correspondence with the author, presented the following observations and information:

"Very interesting account! The exaggerations could be partly due to using candles, but also may represent the fact that not much was known about the cave before the 1877 Wheeler mapping expedition. It includes a reference to rumors of a cave that 'abounded in beautiful grottoes, showy chambers, crystal stalactites, and great lakes in whose waters eyeless fish were found'. The facts, according to their report were 'This cave was thoroughly explored and surveyed, and (we) were not a little disappointed to find its glowing mysteries dwindle into a very commonplace limestone cave'.

"So my guess is that they (Upson) did not go very far (just the Main Corridor) and were just taking their time enjoying the experience. Since the letter was dated 1872, this was before the Wheeler Expedition survey so the 'rumors' were still not influenced by engineering measurements. There is sort of a running joke among cavers regarding the 'cave mile'. Many cavers, especially those of limited experience, overstate underground distances significantly and consistently. In an article in the Eddy County Current in 1897, the location of the old 'canoe' (really Conrad's rowboat) is said to be two miles from the entrance. It is really about 3,000 feet or 0.6 miles. In 1898, W. H. Tuttle described a place where'At a distance of four miles the water covers the entire bottom of the cave to a depth of from two to three feet.' This is probably at Inscription Rock also so 0.6 miles was perceived as 4 miles.

"The 'lofty dome' was what we call the Devil's Backbone Dome which is a nominal 15 minute walk from the entrance under dry conditions. This dome almost fits the description in the old text (it would be considered lofty, especially by the light of a candle) except it is not far from the entrance...

"Since the Fort was only 3 miles away, and built as early as 1855, it is very likely that the cave was discovered sometime in the early 1800's by settlers, but that discovery is not recorded if it did happen. The first recorded mention of the cave is through its discovery during a short chase by the Cavalry of a band of Indians (Apaches) near the Fort. ..., and this has long been the accepted story of the cave's discovery. ... These troops were under orders to kill all Indian men of the Apache tribe 'whenever and wherever you find them'. In the process of carrying out these orders, one patrol from the Fort chased a band several miles cross country, and just as they came up over a hill, they saw the ponies of the Indians near the bottom of a canyon. The ponies were hitched to a clump of trees, and the Indians were nowhere in sight. The soldiers supposedly followed the moccasin tracks to the edge of a large sinkhole which sounds like an excellent trick to make the soldiers think that they had entered the cave. Thinking that they had the Indians trapped, the soldiers made camp, and waited for them to come back out. Several days later, the Indians crept up to their horses, and rode away in a cloud of dust, leaving the soldiers behind. There has long been speculation of the existence of another entrance which the Apaches used to escape from the cave."

(John Corcoran in emails, 2007)

The real question though is whether or not the Indians ever entered the cave in the first place. John Corcoran believes that this is not the most likely occurrence, since there is no evidence of a second entrance that people could escape through. At any rate, this was the "beginning of a large number of references to the cave that soon became the favorite place for 'adventures' among the local people."

One of those locals was a young man by the name of Willett White who grew up at Fort Stanton during the Public Health Years. . He knew every nook and cranny of the vast countryside including the cave. He wrote of his adventures in the cave…

"The Fort Stanton cave was an attraction that we could not resist. There were stories of ambulatory patients from the hospital that became lost in the cave and had to be rescued. Also, there was the story of Indians who had been pursued by the Army and who had sought refuge in the cave. The Army camped outside the cave but the Indians never came out. The Army even built a punt to better travel the water in the cave, but never found the Indians (Apaches). Later, the Apaches were reported to be in the Capitan Mountains.

"Four of us boys spent a whole day in the cave and believe that we were the farthest into the cave of anyone at that time (1920's). We believe we found a dam that the Indians made to back water up to prevent being followed. We signed our initials above the water line, about five to six feet above the floor of the cave, and came out. We had gone in during a drought period and later there was water running in the cave and we were unable to pass a low place that was now underwater. The boys involved were Bob Hobbs (son of the Butcher), Buddy Patterson (nephew of the Assistant Engineer), Terry (I forget his last name) and myself. " (An article covering this event was published in a spelunkers magazine, SOUTHWESTERN CAVERS July-August 1996).

…and The reference to *"The Army even built a punt to better travel the water in the cave…"* is likely referring to the 1872 wooden boat. Another excerpt from John Corcoran's article continues:

"In 1872, a Captain Conrad and a Lieutenant Boyd had a small boat built by J. D. Swan, a carpenter at the Fort, for the purpose of exploring the partially flooded north end of the Main Corridor. The two men apparently spent several days exploring the Main Corridor, and told of exploring to a distance of 8 miles and finding

a beautiful waterfall that dropped into unknown depths. It was also reported that Captain Conrad was in the boat when it capsized and nearly drowned. Pieces of this boat can still be seen near Inscription Rock and in the Main Corridor (Conrad's Branch) as far as Sewer Pipe...

"Today, what we believe happened, is that the boat was carried as far as Inscription Rock (this would have been quite a hard task!) in the Main Corridor, but when they tried to step into the boat, it capsized and sank. The boards of which the boat was built are still seen under a low shelf at Inscription Rock and a few pieces are scattered along the Main Corridor to the north. We do not know where the "Beautiful waterfall" was, but when water levels are not too high, there are a number of places where the water flows over rocks in small waterfalls along the Main Corridor north of Inscription Rock."

(John Corcoran, numerous emails, 2007, 2008 regarding Cave history and folk lore.)

Courtesy Fort Stanton 'Kids' Photo Archives.

Willett White visited the cave many times during his growing up years at the Fort.

From the time of the Dragoon venture into the cave in 1855 let us fast forward about 20 years and, according to John Corcoran, *"a U.S. mapping team, led by Lt. Charles C. Morrison of the Sixth Cavalry in 1877, as part of the Congressionally mandated U.S. GEOGRAPHICAL EXPLORATIONS WEST OF THE 100TH MERIDIAN"* continued their mapping of Fort Stanton cave. Better known as the Wheeler Expedition, these men mapped more than two miles of the cave. Eighty years later a total of more than seven miles had been mapped and surveyed by various exploration parties. Presently, (2010) 14.75 miles of passages are known, and even more of this complex of caves and underground passages continue to entice and entrance explorers. The spectacular Snowy River passage discovery upped the ante not only because of its incredible beauty, but because of the significant scientific and hydrological information that may yet be determined as the research teams continue their investigations.

Dr. Penny Boston and others who are responsible for mapping and exploring this newest dazzling discovery, like John Corcoran, Mike

Courtesy BLM/John Cochran

After a two year planning process led by the Roswell BLM office more than 7.25 miles of new passages have been located, and no one knows what lies ahead.

Bilbo, John Cochran, Lloyd Swartz, John McLean, the Peermans, Donald G. Davis, Wayne Walker, Pete Lindsley, and many others note that the breakthrough came in 2001 although the exploration of the region's underworld has been an ongoing series of expeditions starting in the early 1970's. Snowy River was a stunning and unexpected find that literally took their breath away! An awesome ribbon of snowy white calcite has yet to be fully explored or mapped.

Though very different in context, it is still considered a part of the much better known Fort Stanton Cave complex. This progression of use from natural to ceremonial to awe struck humans of the 19th- 21st centuries continues to amaze researchers. Although people of the past used the cave, they never knew about Snowy River. This is a first.

Snowy River passage is the longest calcite pool formation (approximately 4.6 miles and still going) ever found in the world, according to Mike Bilbo *"Cave scientists are coming to our region from all over the world to pay homage and to continue the necessary work to document and map, to wonder at its beauty, and to help prepare management plans for the future."*

Even the names given to special formations or passages can illuminate the exhilaration and the frustration of the often dangerous work of locating a new and unusual passage that could have unique scientific and aesthetic ramifications for our world. *Roaring Gate, The Metro, Snowy River, Escalator, Lincoln's Bath Tub, Round House, Painted Cliffs, Turtle Junction, Crystal Crawl, The Devil's Backbone, Twenty Steps, Bridal Chamber, Disappointment Dome, and Starry Nights Passage* indicate the robust, the delicate, and the unknown. And the brave lads who first entered the cave in 1855? Wonder what their thoughts would be today if they could make a trip "back to the future?"

Owing to the pristine nature of the calcite river passage neither could the cavers walk on that sparkling river; otherwise their muddy clothing and boots would have quickly destroyed that which they had found. They eventually developed a technique to change into clean clothes and white-soled shoes to preserve the delicate, white river. However, for the moment they backed off and decided to return at a later time to continue their explorations into the unknown realm beyond Snowy River.

Tom Gould exploring cave, c 1955.

Courtesy Tom Gould.

Courtesy BLM/John Cochran

Donald G. Davis, Fort Stanton Cave Study Project, Wheeler Hall, 2009.

That is one of the most exciting parts about this passage, the unknown. The biological data is still unfolding. For example, Dr. Boston claims that there are many novel species unfamiliar to scientists… *"Each cave that we are studying yields up new sets of hundreds to thousands of new organisms. Such untapped biological wealth is ripe for exploration seeking sources of new pharmaceuticals, industrial gems like ….enzymes, …and the microbial role in the very production of economically significant low temperature ores such as uranium, gold, copper, manganese …Geo-chemistry and isotopic data from both the sparkling calcite and other materials in the cave are (also) of great interest."*

John Corcoran.

Photo by David L. Tremblay

"Climatological and hydrological data can also be garnered from this discovery." noted team leader, John Corcoran, "and mapping is a challenge. While deep within the earth, the explorers communicate up to the surface with low frequency CB type radios equipped with a PVC framed loop antenna. These radios are used in sequence to help establish location within the cave and are useful for both mapping and safety factors."

Corcoran has been at this adventure for over 40 years and has seen hi-tech methods improve their research capabilities. These men and women, like the Dragoons and Morrison expeditions before them, are making history happen and it is exhilarating to be a part of this scenario.

Courtesy Tom Gould

Tom Gould, c 1955, Wheeler Hall.

It has been a continuing labor of love for the members of the Fort Stanton Cave Study Project. Despite the hazards, these dedicated men and women have realized the challenge and the opportunity as they crawled and walked, meandered and sometimes by missing a step, free fell along a steep slope or missed a step and tumbled forward against harsh and jagged rock formations in their quest for knowledge. Their photographs also bring to the rest of us the beauty, the drama and the significance of their research, but nothing is like experiencing the adventure of caving first hand.

This honeycomb of caverns and underground rivers beneath the Fort Stanton lands also has had former Senator Pete Domenici and Senator Jeff Bingaman working closely together to create the Fort Stanton-Snowy River National Cave Conservation Area. Careful management practices and plans are being devised to protect the site and will be a plus for our region.

Perhaps one day in the near future, segments of the new discovery can be revealed to the public. The BLM is already working on a "virtual tour" of the cave passages. It is a rare privilege to know that we, here in Lincoln County, have such breath taking treasures with many attributes that may benefit our world beyond measure. However, for now, the new discoveries are adding yet another dimension to the fascinating history of Fort Stanton Cave and Fort Stanton State Monument.

For the first time, these walls of stone
and man meet face to face;
the silence broken, the darkness set aside.
Glitering secrets revealed, as lights fill the space.

Poem by Sigurd (Sandy) Szerlip (1969), inspired by the Lincoln Caverns breakthrough, and copied from the original register.

SIGNIFICANT DATES FOR FORT STANTON CAVE:

(Courtesy Mike Bilbo, John Corcoran, Tom Gould and Lynda Sánchez)

The Bureau of Land Management and the Fort Stanton Cave Study project continue their research and documentation of cultural and natural history of this unique cave. And better yet, an excellent and current update from Ms. Theresa Herrera, Public Affairs Specialist with the BLM brought very good news March 25, 2009… *"Bill passed today and President is expected to sign…Fort Stanton (cave) is finally protected."*

This protection comes at a significant time and means that funding, staff and the research will continue for many years to come.

In terms of dates, the following some but not all that are are important to cave history. Eighteen inscription sites have been discovered and no doubt others will be located.

March, 1855: Earliest known date for non-Native American exploration. Approximately one half mile into the cave are the names of Privates Emil Fritz, Victor Brown, Joseph Myers, Horace Belknop, John Cherry, Louis Loeslie and John Lepsey.

1872 Ash Upson's account of exploring the Cave include colorful descriptions yet to be documented.

August 21, 1877: Members of the Fifth U. S. Infantry in conjunction with the Wheeler mapping Expedition inscribe their names.

August 26, 1887: Inscriptions of 6th and 8th Cavalry from Fort Bayard are included in the growing list of cave explorers.

April 9, 1891: "Three Days and Nights Spent Among the Wonders of a Midnight World" is published in the Great Divide Newspaper of Colorado Springs. Some names from this foray into the cave are written on cave walls and recorded by Bilbo's team.

For members of the public with an interest in visiting Fort Stanton Cave or obtaining more information for group slide presentations about Snowy River, one may contact: Bureau of Land Management, Roswell Field Office, 2909 West 2nd Street, (505) 627-0272. Or Mr. Mike Bilbo, 575-354-5034 (www.NM.BLM.Gov/recreation/roswell/fortstantonacec.htm) In addition to the special use cave permit, you will be required to provide your own caving equipment.

c 1900 Snowflake Passage.

1950's and 1960's: Intrepid young men, mostly with the Boy Scouts or just locals who were adventuresome and curious continue to enter the cave for day journeys or sometimes even longer. Little did these fellows realize that this special "world beneath their feet" would one day be a famous National Cave Area. (John Thomas, Stan Thomas, Aron and Samuel Sánchez, and others remember their days inside the dark cave.) Mr. Tom (Skip) Gould, who grew up at Fort Stanton, has been an invaluable source of information for the cave scientists. He describes spending days in the cave during the 1950's and tells author Sánchez about this exploration with other boys from Fort Stanton. He details entering the cave with flash lights, candles, string and a sack lunch along with a whole heck of a lot of guts. He provides color slides to corroborate his visits. Sánchez passes this valuable data on to the Fort Stanton Cave researchers. This allows them to compare conditions of 60 years ago with what exists in certain passages today.

1958: ***"New Section"*** (First major scenic discovery since 1800's)

August 9, 1969: Lincoln Cavern; Bat Cave Extension 8/23/1969

February 14, 1971: Don Sawyer Memorial Hall.

1975: Fort Stanton Cave is designated as a National Natural Landmark and a major exploration team led by Cave Specialist Mike Bilbo begins recording known sites and looking for additional inscriptions.

September 2001 to present: Snowy River Complex (consisting of innumerable passages with several yet to be named) is discovered. Mapping and surveys continue.

March 25, 2009: a new era begins with President Obama's signature designating the area a National Cave Conservation Area, and "will provide permanent protection to the Snowy River formation and protect it for future research and educational purposes."
(News Release, March 25, 2009, Office of Senator Jeff Bingaman)

FORT STANTON CAVE IS, WITH MORE THAN 14.75 MILES OF MAPPED PASSAGES, THE THIRD LONGEST CAVE IN NEW MEXICO. IT IS OPEN FOR RECREATIONAL USE BY PERMIT FROM APRIL 15 TO NOVEMBER 1. THE CAVE IS CLOSED TO THE PUBLIC DURING THE WINTER MONTHS TO PROTECT HIBERNATING BATS.

Courtesy BLM/Mike Bilbo

Author Lynda Sánchez "Experiencing the adventure of caving, first hand!"

SOURCES AND RECOMMENDED READING:

Bilbo, Mike, (many interviews and special assistance to the author)

Bilbo, Mike, Paper entitled: "The First Dragoons and Fort Stanton Cave"

Corcoran, John (many emails and interviews 2007-2009 to author)

Upson, Ash, Descriptions of Cave from Eve Ball papers. Courtesy Lynda Sánchez

"On the Web"

National Speleological Society:
http://www.caves.org/

Commercial Cave Conservation Association:
http://cavern.org/hrc/hrchome.php

Lechuguilla Cave:
http://www.nps.gov/cave/naturescience/lechuguilla_cave.htm
http://www.extremescience.com/LechuguillaCave.htm

Krubera, World's Deepest Cave:
http://ngm.nationalgeographic.com/ngm/0505/feature4/multimedia.html

(below) Kat Rix on Snowy River.

"I hid my glory too long ... A snowy path of wonder ..." from the BLM poster

Courtesy John Ganter

Courtesy BLM

Linda Rundell, Director of BLM New Mexico, and Lieutenant Governor Diane Denish proudly attend the designation of the Fort Stanton National Cave area, 2009.

Courtesy BLM/Mike Bilbo

Mike Bilbo inside the Circle Room of the Fort Stanton Cave.

Photo by David L. Tremblay

Senior Officer's Quarters, Fort Stanton Quadrangle. Used as staff quarters during the Cavalry and Public Health years. Gen. Black Jack Pershing also stayed here.

Photo by David L. Tremblay

Senior Officer's Quarters.

Chapter 7

A Line in the Sand and a New Mission

"ALL CALCULATIONS BASED ON EXPERIENCE ELSEWHERE, FAIL IN NEW MEXICO!"

(GOV. LEW WALLACE, NEW MEXICO TERRITORIAL GOVERNOR 1878-1881)

"Developing a large subdivision with apartments, town houses and single-family homes around historic Fort Stanton would desecrate the history of the Mescalero Apache traditional homelands and harm an important landmark," said former Mescalero President Mark Chino.

(*THE RUIDOSO NEWS*, August 30, 2005)

"We aren't going to have much of the West left if we don't draw a line in the sand…This is where I drew the line!"

(Lynda A. Sánchez in interview with *TRUE WEST MAGAZINE*, 2007)

"...Fort Stanton is imbued with the stories of countless souls who have left their footprints on this sacred ground. In the shadow of Sierra Blanca with its beautiful setting along the Rio Bonito, the Fort has been called the 'Crown Jewel' of cavalry forts in the Southwest. Please sign the legislation now to designate it a State Monument."

(From letter to Gov. Richardson, Joseph H. Arcure, July 12, 2007)

"There is much more to Fort Stanton than just the Quadrangle...it is the entire 22,000 plus acres that surround this awesome place; it is the calm of the Merchant Marine Cemetery; it is the whisper of bugles long silenced; it is the land of the Mescalero, the Hispanic and Anglo pioneers; it is our unique heritage, and we finally have that mantle of protection..."

(Lynda A. Sánchez, Interview, *Roswell Daily Record*, May, 2007)

"...Fort Stanton is an awesome historical place and its integrity and water rights need to be protected. Please sign the proclamation so that the State Monuments' Management team can begin its work."

(Richard and Mary Samuels, July 12, 2007)

Reenactor Larry Pope.

Photo by Joseph Arcure

"I'm pleased to inform you that Governor Richardson has signed a proclamation designating Fort Stanton a State Monument. The official presentation will take place August 11, 2007 at Fort Stanton."

(Lieutenant Governor Diane Denish to Dr. Noel Pugach in response to questions as to why the continued delay in signing the proclamation...July 30, 2007)

"We must be patient, Monuments are forever and we must do it right."

(New Mexico State Monuments' Director, Ernesto Ortega, August 11, 2007)

"Our docents and volunteers are the unsung heroes..."

(Charlotte Rowe, Fort Stanton Advocate, March, 2009)

"My heartfelt thanks to you and others there for all the work you have put in on 'savin the ole fort.' Congratulations on achieving its State Monument status."

(Marc Simmons, *New Mexico's Historian*, March 31, 2009)

Photo by David L. Tremblay

Fort Stanton Nurses Quarters Building in front of Majestic Sierra Blanca.

Considered by many to be the crown jewel of military forts in the Southwest, Fort Stanton is listed on the national and state register of historic sites and has been recommended to become a National Historic Landmark or perhaps even a National Monument or Park some time in the future. However, the August, 2007 Proclamation declaring Fort Stanton as a New Mexico State Monument was a momentous and happy occasion for the Fort's supporters. This date provided the mantle of protection that will be needed for future preservation efforts.

The quotes noted within are but a sampling of hundreds of letters, calls, interviews and emails that were sent to the Governor's office, his cabinet officials, newspapers and every other place in between. The delays, the half truths, and the poorly conceived Fort Stanton Re-Development Master Plan created so much anger that an unbelievable chorus of Fort advocates world wide clamored for preservation, not a subdivision, to save the fort. That plan was like killing the patient to save the patient.

Photo by Joseph Arcure

When the initial document came out almost everyone was shocked at its lack of imagination and alternatives for a specific management plan. When the Fort Stanton Development Commission accepted this plan and the suggested 600 plus unit subdivision or so called "planned" community was out in black and white, something had to be done immediately to prevent further consideration. And it was. Two of the seven member commission said NO, but only one realized that a line in the sand had to be drawn. That line was made, the battle front had been selected, and the confrontation began, ending three years later with the Fort established as New Mexico's newest State Monument.

"You never argue with a woman carrying a sword," Quipped James Sánchez, husband of author Sánchez after she had just received the Fort Stanton Inc. fourth annual preservation award, a beautiful Cavalry sabre, August 2007.

Long story short, the voice of the people prevailed here and a group of hard working and hard hitting locals voiced their opinions loud and clear to Santa Fe and beyond. Finally the Governor's office representative wrote to author Sánchez: *"Lynda, we hear you loud and clear."* In addition there had been a petition of over 1000 valid signatures objecting to this abominable plan and some of the letters mailed to the Governor and various dignitaries all decried the fact that such a magnificent site would be desecrated with yet another subdivision in an already arid land that has major water problems.

Courtesy New Mexico State Monuments

Bill Hume, Office of the Governor, getting ready to read the Proclamation, standing with Denish, Stuart Ashman, Cabinet Secretary of Cultural Affairs and Earl Pittman, President Fort Stanton, Inc.

Commission members who had been "for" the plan finally relented, stating that it was just a "big misunderstanding." Be that as it may, individuals who should have known better finally bowed to the voice of the people. The Commission had no more meetings and Governor Richardson declared the site a New Mexico State Monument. In August of 2007, the Proclamation was read at a very lively ***Fort Stanton Live*** presentation. Lieutenant Governor Diane Denish attended as did several special supporters from Santa Fe (Bill Hume, John Garcia, Stuart Ashman, and Ernesto Ortega among others).

Should the developers rear their ugly heads in the future, there will no doubt be another major battle, but for now we have a respectable mantle of protection and we are surrounded by 25,000 plus acres of BLM lands (ACEC or Area of Critical Environmental Concern and the newly designated National Conservation Cave Area) thus ensuring that for the next 150 years the area will remain much as it is today, and much as it was 150 years ago. This is a

Courtesy New Mexico State Monuments

(Left to Right) Former State Representative Dub Williams, Lieutenant Governor Diane Denish, Cabinet Secretary for Veterans' Services, John Garcia, and Author Lynda Sánchez celebrating the reading of the proclamation.

GREAT SEAL OF THE STATE OF NEW MEXICO 1912

STATE OF NEW MEXICO **EXECUTIVE OFFICE** SANTA FE, N

Proclamation

WHEREAS, *Fort Stanton was established as a military post in Territorial New Mexico in 1855, was named after Captain Henry W. Stanton, 1st U.S. Dragoons, and was an active military post until 1896; and*

WHEREAS, *Fort Stanton was central to frontier development and was associated with the major 19th Century upheavals of the region including the Apache Wars, the Civil War in New Mexico, and the Lincoln County War; and*

WHEREAS, *Fort Stanton was the first assigned post for Second Lieutenant John J. Pershing in 1887, later referred to as Black Jack Pershing, Commander of the American Expeditionary Force in World War I; and*

WHEREAS, *Fort Stanton served as a Marine Hospital for tuberculosis patients from 1899 to 1953, and was the hospital for the region during the 1930's and 40's providing emergency care to local citizens and people participating in the near-by Civilian Conservation Corps and National Youth Administration camps; and*

WHEREAS, *the CCC camp at Fort Stanton was expanded in 1941 to accommodate a World War II internment camp, primarily for German internees, and was the first internment camp of the War; and*

WHEREAS, *Fort Stanton uniquely represents the era of the 19th century frontier military post in New Mexico and the 20th century era of the tuberculosis sanitarium; and*

WHEREAS, *Fort Stanton is the only intact New Mexico fort originally built before the Civil War and is one of the Southwest's most significant and best preserved; and*

WHEREAS, *Fort Stanton was officially recognized for its historic and architectural integrity through designation on the State Register of Cultural Properties in 1969 and National Register of Historic Places in 1973 and 2000;*

NOW, THEREFORE *I, Bill Richardson, Governor of the State of New Mexico by virtue of the authority vested in me by the Cultural Properties Act, Section 18-6-6 NMSA 1978, do publicly proclaim that there is here created and set aside a State Monument to be know as Fort Stanton State Monument covering the portion of land defined as the historic district by the National Register of Historic Places.*

Attest:

Mary Herrera
Mary Herrera
Secretary of State

Done at the Executive Office this 8th day of August, 2007.

Witness my hand and the Great Seal of the Great State of New Mexico

Bill Richardson
Bill Richardson
Governor

Courtesy

Robert M. Utley.

true salute and legacy for all those who came before and generations yet unborn.

We have come full circle and as the old timer stated, "tell her true". The challenge will be in completing the circle, in being true to the cavalcade of historic characters, events and acknowledging the architectural integrity and natural beauty of the area.

Additionally Mr. Bob Utley, former National Park Service historian and author of many scholarly books about the unique history of the West, has offered a few comments of his own regarding the prospects for the site. The slant is different but the idea is the same…the Fort merits much more than what it has achieved so far. These thoughts are mentioned below as possibly part of a total package sometime in the not so distant future.

Fort Stanton and its Neighbors

by Robert M. Utley

In the 1960's from my perch as Chief Historian of the National Park Service in Washington D. C. I clearly understood the significance of Fort Stanton, New Mexico. For one thing, I had come to Washington from Santa Fe, New Mexico, where I had served as regional historian of the Southwest Region for seven years. I had visited the fort in my travels surveying historic places throughout the region.

More important, in my spare time I was working on a two-volume history of the U. S. Army and the Indian. The role of Kit Carson in the war against the Mescalero Apaches of Sierra Blanca alone stamped the post with significance in American history. Years later, when I undertook a history of the Lincoln County War and a biography of Billy the Kid, my initial evaluation was greatly strengthened. Fort Stanton played a major part in those dramatic events…
No question remains in my mind about its national significance. I applaud the efforts of those who have labored so long and hard to

bring the fort back to life and to attract the interest of the state in further recognition and improvement. Thanks to the efforts of Lynda and many citizens and state agencies, Fort Stanton has regained much of the integrity that it lacked more than forty years ago…

But my thinking has long since expanded beyond the future of Fort Stanton alone. It is linked in history and locality with other historic places–the town of Lincoln, San Patricio, the Chisum Ranch, Fort Sumner, the Mescalero Apache agency, and others.

Troops from Fort Stanton were sporadically involved in the unfolding events of the Lincoln County War, intervened decisively in the critical battle for Lincoln in July 1878, and were part of the continuing saga of Billy the Kid. Supply contracts for both Fort Stanton and the Mescalero Agency underlay much of the hostilities of the Lincoln County War, and troops intervened in several wartime instances that occurred at the Mescalero Agency. Not linked to Fort Stanton were battles at San Patricio and the Chisum Ranch, and Fort Sumner furnished the setting for the final moments in the life of Billy the Kid. The Mescalero Agency was the place where the celebrated gun battle and killing of Buckshot Roberts occurred. Beyond its association with the Lincoln County War, Fort Stanton not only achieved importance in Kit Carson's campaigns against the Mescaleros, but took an active part in the warfare with these and other Apaches until the surrender of Geronimo in 1886.

All these events and places, so clearly associated in history and place, fully entitles them for inclusion in a National Historic Park. National Monuments and National Historic Sites are single entities. A National Historic Park bundles a number of related sites into a single park. They are not all owned and operated by the federal government but joined by cooperative agreements that specify how they are to be managed and interpreted.

Fort Stanton Post Office, representing decades of history.

Photo by David L. Tremblay

Photo by David L. Tremblay
Photo by David L. Tremblay

The National Park Service customarily operates a visitor center to interpret the entire story, and the individual units interpret their stories and manage their resources consistently with the criteria set forth in the agreements.

Fort Stanton is plainly of national significance on its own merits and qualifies for National Monument or National Historic Site status. But it deserves more–as a vital element in a National Historic Park.

Such a park requires an act of Congress. It rarely materializes without the support of the local member of Congress, the governor, and the state's senatorial delegation. Such support is not forthcoming without the active labor of local citizens and organizations working toward the common goal. The part of New Mexico that includes these and other sites has demonstrated in the achievements of such people at Fort Stanton that it could advance such a project.

"Are the people interested and willing?"

Bob Utley has asked an intriguing question. Are the people ready and willing to try for this status? The idea has merit and all we can say at this point in time is that the mission to preserve the Fort has taken a giant leap forward, and perhaps at a later date, when the dust is settled, we can consider this proposition. Perhaps Mr. Utley's vision will materialize in the not too distant future.

More than 200 people attended the grand opening of the Fort Stanton Exhibit held in the Hubbard Museum of the American West, January 2009.

Photo by David L. Tremblay

Mr. Jay Smith, Director of the *Hubbard Museum of the American West* and member of the board of the *Lincoln County Historical Society* offered some thoughts and ideas regarding the potential, the challenge and the opportunity of Fort Stanton. His perspective as a museum man is also of great importance and sheds a different light on some topics. Jay and his staff stepped up to the plate and actually put together an exciting exhibit about the fort. In other words, they put their money where their mouth was! Everyone was thrilled with the fine exhibition and will hopefully be inspired to recognize what we have to accomplish in the future.

Photo by David L. Tremblay

Jay Smith, Director, Hubbard Museum of the American West, welcomes visitors to the grand opening of the Fort Stanton exhibit.

"*In January of 2008*, the *Hubbard Museum of the American West* committed to working on an exhibit about the history of Fort Stanton. I had no idea that my standard approach to working on museum projects would be tested. You see, I come from the school that suggests historians and museum professionals should strive to keep some objective distance between themselves and the topic at hand. As I see it, our job as museum professionals is to provide balanced historical interpretation; "neither an advocate nor a defender be" as one of my former professors so eloquently stated many years ago. While true objectivity is nearly impossible, I never wanted to be the type of museum director who sought deliberately to change the mind or opinion of a visitor. I have always felt that it was not my job to present history in a manner which seeks to convert a visitor towards seeing and interpreting history the way I do. Rather, it is important to present what we know, admit what we do not know, and make sure that our visitors understand what is fact and what is speculation – and hopefully present it all in a manner that is exciting or compelling enough to help our guests draw their own conclusions.

I still feel that way.

However, after working for nearly a year on the Fort Stanton exhibit that opened in January 2009, I found that I needed to make an exception, if for just this once, and be an advocate. Though I had been to Fort Stanton before, walking the grounds with Lynda Sánchez, hearing her passion for the site, and seeing it with renewed interest sparked by months of research brought me to the realization that ***I was looking at a place that deserved greater public and private attention.*** It does not take the imagination of a writer to "see" the past at Fort Stanton, the parade grounds, the officer's quarters, the stables, the administration building, the laundresses houses, all speak to the military and hospital periods of occupation. A quick jump across the Bonito River (when it isn't running) takes you directly to the World War II Internment camp and transports you back to a period of human history that must be remembered in all its manifestations. If you listen closely to the winds howling off the beautiful Capitan Mountains in the distance on any summer day, you can hear the echoes of baseball games, rodeos and workers coming in from the fields when Fort Stanton was a marine hospital. Read any account from the people who once lived at Fort Stanton hospital, close your eyes, and on a crisp autumn afternoon, you can see it as if you were there.

That is the point I think. The Fort is already a Museum worthy of public support and it has the potential to be an attraction that pays educational, cultural, and social dividends for generations to come! Fort Stanton is full of history. Everything about it and around it screams out for recognition on some level or another. It is full of compelling stories. Many are courageous and compassionate. Fort Stanton is a place that was at the center of so many societal changes and emerging needs that it is difficult to realize that the site was already 70 years old in 1925 when it underwent its third large-scale renovation. At each step in its long journey toward the future, the buildings and grounds met some unfulfilled need within contemporary society, leaving behind the echoes and whispers of every human emotion ranging from tremendous sadness to great joy. There is much to learn from the site. Its rich heritage deserves and demands to be shared with future generations.

Few sites can boast of being a fort, a hospital and a prison camp; and fewer still can boast of being the site of native archeological ruins, a

Photo by Joseph Arcure

Buttons and a bottle - Great 'finds' from a recent archaeological survey.

historical cemetery, and a mysterious cave still clamoring for mapping. Of course, today, the conversation has turned to tourism, and we welcome the New Mexico Department of Cultural Affairs and thank them for their commitment to long-term preservation. I can see future generations of kids taking field trips on the weekdays and begging their parents to take them to Fort Stanton again on the weekend to see the live, costumed interpretation. What an opportunity for an entrepreneurial public historian!

Let me be very clear about one thing: I am not of the opinion that all places where historical events took place should be preserved,

Photos by David L. Tremblay

Displays at the Fort Stanton Exhibit, Hubbard Museum of the American West, January 2009.

Photos by David L. Tremblay

or that all "battle" sites should be recreated, or that every home of the famous or infamous demands the attention, time and money it takes to make that leap into the realm of heritage tourism. Nor do I believe, however, that history should be preserved solely due to its association with the famous, infamous, or simply because it is "old". In my estimation, there must be something compelling about the place, person, or event that helps us understand that this should not be forgotten no matter what might transpire in the years afterward.

If history is about who we were, and our personal and collective history informs who we are, then what we save says a lot about our values as a people and a society. What we choose to allow to disappear into the mists of a forgotten past also says a lot about who we are – as well as who we wish to be. ***Fort Stanton has the capacity to bring to life a broad sweep of local, state, and national history in a way that few places can.*** It has all of the elements necessary to provide a fantastic experience to future visitors, if only we as residents of this community, this state, and this nation rally to its support. The infrastructure is already in place, and though there is a great deal of historic preservation work that needs to be performed on the buildings, there are few historic sites in the nation where so much of the original construction is still in place. It will take time, money, and commitment on many levels.

While the exhibit that I worked on has come and gone, the effort of Lynda Sánchez to educate people about the Fort and its opportunities continues unabated. I do not know why, exactly, Fort Stanton came to mean so much to her. Maybe it is in her DNA, maybe it is something that just grew on her, or maybe it is something that she feels so deeply as to be completely inexplicable anymore – it just is. I do know this much, however: Her passion is infectious, her storytelling is compelling and her love of Fort Stanton cannot be questioned. Her willingness to give her own time, energy and considerable talent to the Fort Stanton project is a story unto itself, and I believe that this book is just one of the many bricks in the foundation she has laid toward the realization of the full educational potential of the Fort Stanton State Monument. She has given her heart to the site.

(Left to Right) Tom Stewart, Lincoln County Manager, Ernesto Ortega, Director, New Mexico State Monuments, and Rick Virden, Sheriff of Lincoln County. All great Fort Stanton advocates.

Photo courtesy New Mexico State Monuments.

I know that I speak for many today who are grateful for her actions, and I think I can speak for many future generations who will enjoy the fruits of the labor to which she and many others have given so much, and say, "*Job well done, Lynda. Thank you.*"

And finally, yet another dedicated man of vision has stepped forward to carry the baton regarding Fort Stanton. Mr. Ernesto Ortega, current Director of New Mexico State Monuments also has a varied National Park Service and historic preservation back ground. In that position he has been assigned the guidance and the future development of Fort Stanton. With his steady and experienced hand he has guided the beginning of the process that will last for many years as we forge ahead to open the Fort for the general public. We are most grateful to him and his staff of dedicated planners and interpreters. Although the challenge is great, and the opportunities boundless, we believe Mr. Ortega will pilot this magnificent ship, this wonderful example of our Western American experience to a safe harbor. We will see this come to fruition at the Fort's grand opening that will no doubt garner national and international acclaim for New Mexico at which point we can all say, "*Mission Accomplished!*"

Deedee Tremblay's beautiful smile encompasses the new mission for Fort Stanton State Monument—that is, ensuring that visitors can 'touch the past' with the knowledge that preservation of the Fort will truly be forever!

Deedee Tremblay at the Historic Stables in 2008.

Photos by David L. Tremblay

Photos by David L. Tremblay

"Tell it true."

Re-enactor on porch of Administration Building at a recent Fort Stanton Live event.

Discovered beneath the porch or veranda of the Administration Building, this stunning 1882 "Insignia Keystone" (Cornerstone) was hidden from the world for more that a century. Restoration crews found it during the reconstruction phase. It was possibly carved by a member of the 4th Cavalry. These carved markers were often set out in front of the barracks buildings representing a special unit or color guard. Who knows what other exciting discoveries will be located as restoration continues?

Photo by David L. Tremblay

Photo by David L. Tremblay

Chapter 8

Restoration Portfolio

2010 The second, revised edition of the Fort Stanton book has allowed us the opportunity to include the following photographs of new discoveries and architectural features as well as photo documentation of the restoration/stabilization process. Additionally, photographs have been forth coming from a variety of sources to augment those we have already included in the book. These would consist primarily of the Fort Stanton Cave research and vintage photographs that we hope will enhance the historical records.

In 2008 and 2009, Capital Outlay funding for which the people had so valiantly fought since 2003 was finally beginning to transform some of the Fort buildings that needed it the most. Primarily the Quadrangle (Parade Ground) was the recipient of most of the more than 1.5 million dollars worth of funding for renovation and repair. The historic stables and silos were also beneficiaries of generous private donations and capital outlay funding as well as many volunteer hours.

Photo by David L. Tremblay

Photo by David L. Tremblay

The Administration Building, an 8000 square foot beauty, was to see major work on the columned portal or porch, and its interior. The inside of the structure was in bad need of repair as termite infestation had initially destroyed some of the wooden window frames and several supporting columns. The old wooden floors were replaced, new electrical wiring and plumbing were completed and a series of wonderful exhibit rooms, State Monument offices and public restrooms were installed. A subtle white lime wash replaced the paint on the battered walls of the exterior.

Photo by David L. Tremblay

Photo by David L. Tremblay

In February, another honor was credited to the restoration on site and that was the nomination of the Administration Building for the 2010 New Mexico "Architectural Heritage Preservation Award". There were six co-nominees: Crocker Ltd.; Marty Perrins-Dallman as Project Manager for Property Control Division; New Mexico State Monuments; Fort Stanton, Inc.; Huitt-Zollars; and Edie Cherry of Cherry/See/Reames. On April 15, 2010 the announcement everyone was hoping for came through. The Administration Building had won the coveted architectural award! This is yet another feather in the Fort Stanton cap!

To assist in this nomination process, David Tremblay provided the before and after photos for the application, many of which are visible in this Portfolio. Author, Lynda Sánchez provided editorial assistance, a trove of newspaper and magazine clippings, and a letter of support. Jay Smith, Director of the Hubbard Museum, also provided a letter of support as did Lincoln County Manager, Tom Stewart.

Those who recall the devastation and damage inside and out of this wonderful building will now see a great difference. We congratulate all nominees because each group or person worked toward the eventual restoration and occupation of the structure. In many ways we have come full circle and as Englishman Fred Nolan stated in the introduction to this book, *"A country that turns its back on its past has nothing with which to face the future."* The elegance has returned and the next century has begun with a new face, a new purpose and pride and hope for the future.

Photos by David L. Tremblay

Both the **Senior and Junior Officer's quarters** saw window and door weather proofing, roof repair, a new coat of lime wash, asbestos abatement and clean up and removal of raccoon and squirrel feces and nests, and other structural improvements.

Senior Officer's Quarters. Note replaced shingles on both structures.

Junior Officer's Quarters

Hopefully in the near future these buildings will once again house people instead of "critters" as they have for the last two decades. Removal of rodents and asbestos abatement have been completed.

Photos by David L. Tremblay

The 1877 **Commandant's home** saw the replacement or repair of window lintels (beam across top of window) and stabilization of one corner of the home. Additionally, repair of the rotted fascia corner area in front and chimneys, and replacement of some shingles helped to round out the weatherization and rodent resistance of the structure.

According to the 1988 Williams Co. Preservation Masterplan, this 1877 structure "is the best preserved of all the historic buildings at the fort...the original front door with side lights and transom are still intact. The woodwork, walls, and floors are in excellent condition." (page 34)

Photos by David L. Tremblay

Eagle insignia from the door of the Fort safe

Photos by David L. Tremblay

Historic Stables and Silos

The historic stables that date back almost to the 1855 founding of the Fort now have the walls rebuilt and according to Cornerstones and SHPO regulations. The fascia board has been replaced and new sheet metal (tin) roofing has been installed. The Dairy section has been stabilized and cleaned up and the silos' roofs will soon be completed. Serving both the Cavalry era and that of the U. S. Public Service years accounts for the more than 150 years of existence of these buildings.

Rusting tin roofs could not last one more year of wind and rain damage. In 2009 Capital Outlay funding helped to replace all roofing and repair walls and other features at the stables.

Visitors love this part of the Fort and often spend (on special tours) about an hour photographing the site, the beautiful sandstone walls, and wondering around into the stalls, the Farm Foreman's office, the silos, and then searching the sand stone walls for carvings or dates and initials of the builders.

Photos by David L. Tremblay

Photo by Deedee Tremblay

Photo by David L. Tremblay

Photo by David L. Tremblay

The 1930's silos have become a "signature" symbol of Fort Stanton. Photographer Tremblay, on assignment, climbing to the top photographed these incredibly beautiful structures inside and out to assist in future stabilization and renovation planning.

Photo by David L. Tremblay

Photo Courtesy Fort Stanton 'Kids' Archives

Special design feature on interior of the WPA Nurses' Quarters, (soon to be BLM headquarters). Hand carved corbel and vigas typical of 1940's Pueblo Revival Style.

Photo Courtesy Fort Stanton 'Kids' Archives

Photo by David L. Tremblay

The Bureau of Land Management is negotiating a lease with the State of New Mexico to occupy the former WPA era nurses quarters. They hope to make repairs on the interior of the building as well as painting and renovating the outside. It will become the headquarters for their National Conservation Area offices and serve as a meeting and study site for the Fort Stanton Cave Study Project. (See page 52-53)

A special thanks to Dee Kessler, Interim Monument Manager and her staff; Dr. Richard Reycraft, Cultural Resources Manager for NM State Monuments; Dorothy Victor, State Historic Preservation Office; Project Managers Marty Dallman and Martin Hibbs, from Property Control Division for pushing through all the projects funded by private donations, capital outlay; and a SAT grant (Save America's Treasures/ NPS funding.)

Photographer David L. Tremblay, my partner in this book endeavor, has provided the additional photographic documentation of this inspiring renovation. We now have "happy" buildings and once this impressive cultural resource is opened to the public, we will have completed our responsibility in preserving an awesome site for future generations.

Lynda A. Sánchez
May 2010

Photo by Joseph Arcure

Photo by David L. Tremblay